Aliaksandr Navitski

Scanning field emission microscopy

AF062748

Aliaksandr Navitski

Scanning field emission microscopy

Scanning field emission investigations of structured CNT and MNW cathodes, niobium surfaces and photocathodes

Südwestdeutscher Verlag für Hochschulschriften

Imprint

Any brand names and product names mentioned in this book are subject to trademark, brand or patent protection and are trademarks or registered trademarks of their respective holders. The use of brand names, product names, common names, trade names, product descriptions etc. even without a particular marking in this work is in no way to be construed to mean that such names may be regarded as unrestricted in respect of trademark and brand protection legislation and could thus be used by anyone.

Publisher:
Südwestdeutscher Verlag für Hochschulschriften
is a trademark of
Dodo Books Indian Ocean Ltd., member of the OmniScriptum S.R.L Publishing group
str. A.Russo 15, of. 61, Chisinau-2068, Republic of Moldova Europe
Printed at: see last page
ISBN: 978-3-8381-2489-6

Zugl. / Approved by: Wuppertal, BUW, Diss., 2010

Copyright © Aliaksandr Navitski
Copyright © 2011 Dodo Books Indian Ocean Ltd., member of the OmniScriptum S.R.L Publishing group

Abstract

Scanning field emission (FE) investigations of structured carbon nanotube (CNT) and metallic nanowire (MNW) cathodes, niobium surfaces and photocathodes are reported in this thesis. A novel scanning anode FE microscope (SAFEM) has been developed within this doctoral work. The microscope will be a part of the systematic quality control of freshly prepared photocathodes at DESY. It is designed to achieve dc surface fields of at least 200 MV/m and provides the localization of field emitters with a spatial resolution of about 1 µm. Design, control software, assembly and actual status of the microscope will be presented.

Varying arrays of CNT columns and blocks were fabricated by two different chemical vapor deposition (CVD) methods. The properties of the structured cathodes were measured by FE scanning (FESM) and scanning electron (SEM) microscopy. Well-aligned FE from nearly 100% of the patches at electric field <10 V/µm was observed. High current capabilities of the columns up to mA and stable currents up to 300 (100) µA for pure (TiO_2 coated) CNT blocks, were achieved. Integral FE measurements with luminescence screen (IMLS) and processing under N_2 and O_2 exposures of up to 3×10^{-5} mbar demonstrated rather homogeneous current distribution and long-term stability of the CNT cathodes.

The cathodes containing regular patch arrays of random metallic nanostructures, as an interesting alternative to the CNT, were fabricated by an electrochemical deposition of Au and Pt nanowires (NW) in ion track-etched templates and were systematically investigated with the SEM and FESM. FE with about 90% efficiency was achieved. The current carrying capability of individual patches, however, strongly varied between 40 nA and 90 µA. Actual current limits are caused by heating and successive destruction of NW. Electro-thermal model calculations and the SEM images reveal geometrical constrictions in NW contact region as a reason for the observed current limitations.

Systematic measurements of the surface roughness and local defects on high purity Nb samples by means of optical profilometry, atomic force microscopy and SEM, as well as their contribution to FE as measured by FESM and derived geometrically are reported. Particulates and scratches were identified as potentially stronger field emitters than grain boundaries, round hills and holes. It was shown that the defects with electric field enhancement factor $\beta_E \geq 50$ for XFEL and $\beta_E \geq 20$ for ILC should be completely avoided for successful suppression of a parasitic FE load. Many large pits with crater-like centers and sharp rims found on the surface of real cavities as well as the hills and holes hint for problems with chemical surfaces treatments. They do not cause enhanced FE but have to be considered as sources of quenches and magnetic field limitations.

Contents

Symbols and abbreviations   v

1. **Introduction**   1
2. **Theoretical background of electron field emission (FE)**   3
    2.1. Metallic surfaces with local field enhancement   4
    2.2. Semiconductors and band structure effect   14
    2.3. Cold cathodes with mutual shielding   17
    2.4. Electro-thermal properties of nanostructures   18
3. **Measurement techniques**   21
    3.1. FE scanning microscope (FESM)   21
    3.2. Integral measurement system with luminescent screen (IMLS)   25
    3.3. Novel scanning anode FE microscope (SAFEM)   27
        3.3.1. Actual status of dark current investigations   27
        3.3.2. Construction of the SAFEM   30
    3.4. Surface analysis   36
        3.4.1. Scanning electron microscope (SEM) and EDX   36
        3.4.2. Optical profilometer (OP) and AFM   37
4. **Fabrication and FE results of structured cathodes**   38
    4.1. Arrays of CNT columns and blocks   38
        4.1.1. Preferential CVD synthesis of the CNT columns on Si   39
        4.1.2. Water assisted CVD synthesis of the CNT block arrays   41
        4.1.3. Efficient high-current FE from arrays of CNT columns   43
        4.1.4. FE properties of aligned pure and TiO_2-coated CNT block arrays   50
    4.2. Arrays of gold and platinum nanowire patches   57
        4.2.1. Ion track template technique   58
        4.2.2. FE homogeneity, alignment, and maximum current limits   61
        4.2.3. FESM-SEM correlation study   72
5. **Parasitic FE from Nb surfaces**   75
    5.1. Nb surface preparation   75
        5.1.1. Overview of the cavity fabrication methods   75
        5.1.2. Samples for the surface study   77
        5.1.3. Electropolishing (EP) of Nb   78
        5.1.4. Buffered chemical polishing (BCP) of Nb   79

	5.1.5. High pressure ultra pure water rinsing (HPR)	80
5.2.	Surface roughness and electric field enhancement (EFE)	82
5.3.	Correlated FE/SEM/OP investigations of the flat EP Nb samples	88

6. Summary and outlook   98

Acknowledgements   103

References   105

Appendix A   117

Symbols and abbreviations

3D	Three-dimensional
AFM	Atomic force microscope
BCP	Buffered chemical polishing
CAD	Computer-aided design
CCD	Charge-coupled device
CNT	Carbon nanotube(s)
DC	Direct current
DESY	Deutsches Elektronen Synchrotron
DIC	Dry ice cleaning
E	Electric field
EDX	Energy dispersive x-ray analysis
EFE	Enhanced field emission
E_{acc}	Accelerating electric field
E_{max}	Maximum electric field, maximum effective field
E_{on}	Onset electric field
E_{surf}	Surface electric field
E_0	Macroscopic electric field
EP	Electropolishing
FE	Field emission
FED	Field emission display(s)
FESM	Field emission scanning microscope
Fig.	Figure
FLASH	Free-electron laser in Hamburg
FN	Fowler-Nordheim
GPIB	General purpose interface bus
h	Height
HI	High input
HPR	High pressure ultra pure water rinsing
I	Current
I_{FN}	Fowler-Nordheim FE current
IGP	Ion getter pump
ILC	International linear collider
J	Current density

MIV	Metal – insulator – vacuum
MIM	Metal – insulator – metal
MWNT	Multiwall CNT
LO	Low output
NW	Nanowire
OP	Optical profilometer
PC	Personal computer
PID	Proportional-integral-derivative
ppm	Parts per million
r	Radius
rf	Radio frequency
RP	Rotary pump
RRR	Residual resistivity ratio
S	Effective emitting surface
SEM	Scanning electron microscope
SWNT	Singe wall CNT
t	Time
T	Temperature
TESLA	TeV Superconducting Linear Accelerator
TMP	Turbo molecular pump
TSP	Titan sublimation pump
UHV	Ultra high vacuum
V	Voltage
XFEL	X-ray free electron laser
α	Field correction factor due to anode geometry
$\varnothing$	Diameter
φ	Work function
β	Field enhancement factor
β_{eff}	Effective field enhancement factor
θ	Angle

Introduction

Here I present the work I carried out for my PhD studies on the Physics Department of Bergische University of Wuppertal in field emission (FE) group of Prof. Dr. Günter Müller in period from June 2007 to September 2010. The work is focused on FE investigations of different materials suitable for cold cathodes applications, as well as investigation of origins of a parasitic FE from Nb surfaces and photocathodes.

The thesis consists of five chapters. A short description of each chapter is outlined below.

Chapter 1: is the introduction containing a brief overview of the thesis. Additionally each experimental chapter contains a more detailed introduction and motivation to the field of interest discussed in the chapter.

Chapter 2: is concerned with theory of electron emission from metallic and semiconducting material, description of mutual shielding effect, which is an important issue for cold cathode applications, and given an overview of electro-thermal properties of carbon nanotubes (CNT) and metallic nanowires (MNW).

Chapter 3: presents a brief description of measurement techniques as FE scanning microscope (FESM), integral measurement system with luminescent screen (IMLS), scanning electron microscope (SEM) with energy dispersive x-ray analysis (EDX), optical profilometer (OP), and atomic force microscope (AFM) used during investigations. The chapter includes also a comprehensive description of a novel scanning anode FE microscope (SAFEM) developed during the PhD work. About SAFEM, it was reported also at SRF2009 conference in Berlin [Nav09c].

Chapter 4: describes fabrication and discuses FE results of different structured cathodes based on the CNT and MNW. Several publications in reviewed journals: [Nav09a, Nav10a, Nav10b, Jos10], conference contributions: [Lab07, Lab09, Nav08a, Nav09b, Nav10c, Nav10d, Nav10e, Pru09a, Pru09b, Sol09] and scientific reports [Nav09e, Nav10e] were made based on the results discussed in the chapter.

Chapter 5: gives an overview of Nb surface preparation methods and presents results on surface roughness and electrical field enhancement studies as well as FESM-SEM correlated FE investigations of flat electropolished (EP) Nb samples and is concerned with parasitic FE. The results were presented also as conference contributions [Nav09d, Nav10f].

Chapter 6: gives general summary of the work and some overview for the future studies.

Appendix A: is made up of flowcharts of the LabView [Lab] programs used to control the SAFEM and acquire the measurement data.

A part of the work done during the PhD study was not included in the thesis. Some results of it can be found in the following conference contributions. [Bor10] describes a new apparatus for FE spectroscopy measurements of cold cathodes and gives some first results of measurements of work function of CNT. [Sch10a] is concerned miniaturized FE electron sources for sensor applications based on Si tip arrays. [Sol08a, Sol08b, Sol09] describe process of formation and present FE results of cold cathodes based on CNT synthesized in porous anodic alumina.

2. Theoretical background of electron field emission (FE)

FE is defined as the emission of electrons from the surface of a condensed phase (metal or semiconductor) into another phase, usually a vacuum, under the action of high electrostatic fields [Gom93]. This phenomenon was first reported in 1897 by R. W. Wood [Woo97], who was actually looking for a way to produce very intense x-rays. The theoretical study was started by W. Schottky (1923), who tried to describe the FE by means of electrons thermally excited over a potential barrier at a surface, the width of which is reduced by an applied electric field [Sch23]. The expected dependencies of the emission current on the electric field (I ~ $\sqrt{E}$) and on temperature, however were not observed experimentally [Gos26, Mil26, Pie28]. The first correct mathematical explanation was made by Ralph H. Fowler and Lothar W. Nordheim in 1928 [Fow28], who developed, what is now known as the Fowler-Nordheim (FN) law, also called the FE equation for metal. The phenomenon consists of the quantum mechanical tunneling of electrons through a deformed potential barrier (see Fig. 2.1) of suitable height and thickness at a surface of a metal. Thus, it differs fundamentally from thermionic emission or photoemission, where only electrons are ejected with sufficient energy above the potential barrier.

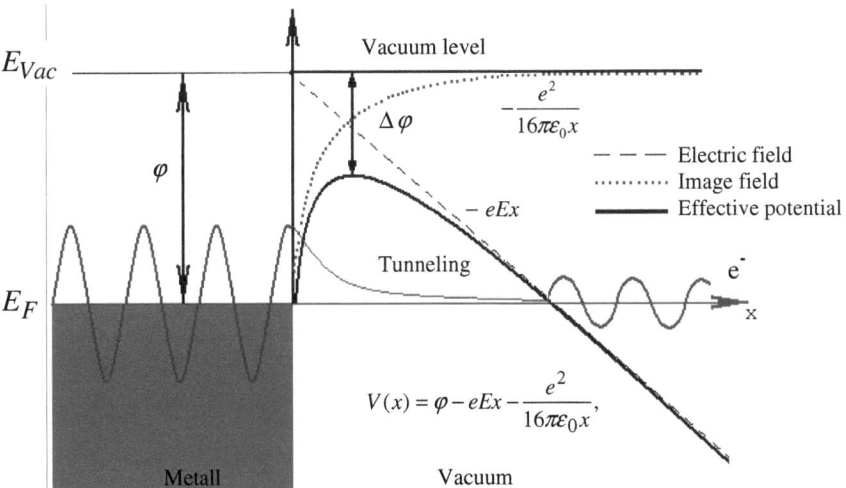

Fig. 2.1: Schematic illustration of the tunneling of electrons represented as a wave function from Fermi level (E_F) via the surface potential barrier (V(x)) at the metal-vacuum boundary.

Electrons in solid are bounded to the core atoms via electrostatic force. The potential barrier as a result of the electrostatic force is called work function (φ). At low temperature, most of electrons have total energy below Fermi level as described by the Sommerfeld free electron model with Fermi-Dirac statistics [Chr88]. The work function of a solid indicates correspondingly minimum energy needed to move an electron from the Fermi level into vacuum via the potential barrier. The form of the potential barrier is, in some distance of the surface, described by the interaction between the electrons located at $+x$ outside the metal and an image charge located at $-x$ inside the metal (mirror charge). Thus, the potential barrier has the form $-e^2/(16\pi\varepsilon_0 x)$ as shown in Fig. 2.1. Very close to the surface, the form of the potential barrier has to change due to the divergence of the mirror charge potential [Sch23]. The detailed form is unknown and in fact irrelevant for our purpose. The external field with strength E in the direction $-x$ is described by a potential $-eEx$ outside the metal. Therefore, the effective potential expressed as [Gom93, For99]:

$$V(x) = \varphi - eEx - \frac{e^2}{16\pi\varepsilon_0 x}, \qquad 2.1$$

where φ is the work function of the metal and ε_0 is the dielectrical constant in vacuum. For example, to decrease the height of the potential barrier by 1 eV, an external field of 0.7 GV/m is necessary. Such strong electric field must be applied to thin down the potential barrier thereby allowing electrons quantum-mechanically tunnel into vacuum. This is called FE because electric field is the main energy source that induces the electron emission.

2.1. Metallic surfaces with local field enhancement

A complete electron FE mechanism from a metal cathode can be illustrated using energy band diagrams of the emitting systems. First, let us consider the planar metal cathode (see Fig. 2.2 (a)). Applying a voltage (V) between anode and cathode creates a uniform electric field E = V/d across the vacuum gap d as shown in Fig. 2.2 (b). If the applied electric field is sufficiently strong, electrons (mostly with energy below the Fermi level) can quantum-mechanically tunnel through the triangle barrier into the vacuum and they are accelerated by electric field until they reach the anode. At the vacuum-metal (anode) interface, electrons collide with the metal and lose their energy as heat.

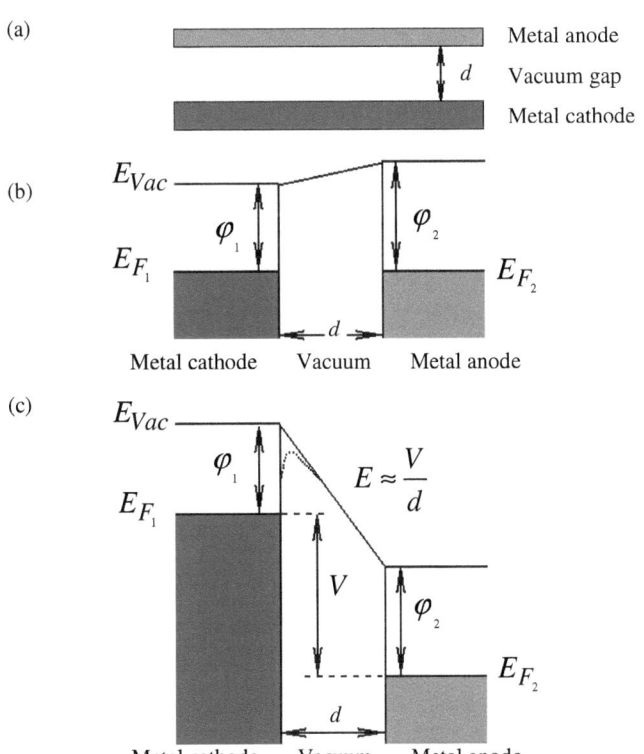

Fig. 2.2: Planar structure of metal cathode-anode (a), energy band diagram of it at thermal equilibrium (b) and energy band diagram under applied voltage (c).

Fowler and Nordheim solved the Schrödinger equation with the Wentzel-Kramer-Brilloin (WKB) approximation in order to calculate the transmission coefficient through the triangular barrier. The FN equation gives the magnitude of the FE current in A as [For99, Goo59, Spi76]:

$$\begin{aligned}I_{FN} &= \frac{e^3}{8\pi h}\frac{SE^2}{\varphi\, t^2(y)}\exp\left(-\frac{4}{3}\frac{(2m)^{1/2}}{e\hbar}\frac{\varphi^{3/2}v(y)}{E}\right)\\ &= 1.54\times 10^{-6}\frac{E^2}{\varphi\, t^2(y)}\exp\left(-6.83\times 10^9\frac{\varphi^{3/2}v(y)}{E}\right)\end{aligned}$$

(2.2)

or in a more simplified form:

$$I_{FN} = A\frac{SE^2}{\varphi\, t^2(y)}\exp\left(-B\frac{\varphi^{3/2}v(y)}{E}\right), \qquad (2.3)$$

where S is parameter interpreted as an effective emission area in m^2, φ is the work function in eV of the emitting surface assumed to be uniform and independent of the external field, E is the field strength in V/m, e is the electron charge in C, m is the electron mass in kg, h is Planck and $\hbar$ is reduced Planck constant. Constants A = 1.54×10^{-6} and B = 6.83×10^9 by field measured in V/m or more practical A = 154 and B = 6830 by V/µm. $t(y)$ and $v(y)$ are tabulated Nordheim functions [Bur53], which depends on the relative reduction of the barrier through the image charge. Often used estimations are $t(y) = 1.1$ and $v(y) = 0.95 - y^2$, which are without image charge correction both set to 1. From the equation, the emission characteristic strongly depends on the work function of the cathode. Material with lower work function gives a higher emission current at a given applied electric field. Considering the absolute value in the exponent of the equation, φ is usually around 4-6 eV for metals, thus $\varphi^{3/2}$ and the exponential factor are approximately 10 and $10^{-11}/E$ respectively. Therefore, an applied field greater than 1 GV/m is required to make any sensible emission measurement from a planar metal cathode [Rot26]. On the other hand, as the work function exponentially affects the emission current density, slight change in work function can cause large current fluctuations. Even thought, in ultra-high vacuum inevitable adsorptions tend to change the emitter's work function, thus cause instability in the FE current. Another problem is that a field emitter suffers from short service life. The high emission current density causes serious joule heating that raises the tip's local current, which increases the atomic mobility at the tip. High electric field at the tip of the emitter tends to form local protrusions through a field build-up. This cycle usually leads to vacuum arching and destroys the emitter structure [Fur05]. Another factor that influences emission is surface roughness. Any surface irregularities show up as areas of intensified emission density. In experiments with real metal cathodes, therefore, the emission current appeared already at fields in the order of 10 MV/m [Alp64, Ben67, Dav68, Nie86, Mah95, Pup96, Hab98, Göh00] instead of GV/m. Therefore, let us consider roughness of a real surface illustrated as sharp cones in figure 2.3. The sharp cone structure is usually recognized as the "Spindt cathode" [Spi68], which has been developed by using various types of metal materials. The sharp cone structure results in non-uniform electric field as illustrated in figure 2.3 (c). The electric field is highest at the tip apex and rapidly decreases outward to the anode as shown by numerical simulations (Fig. 2.4).

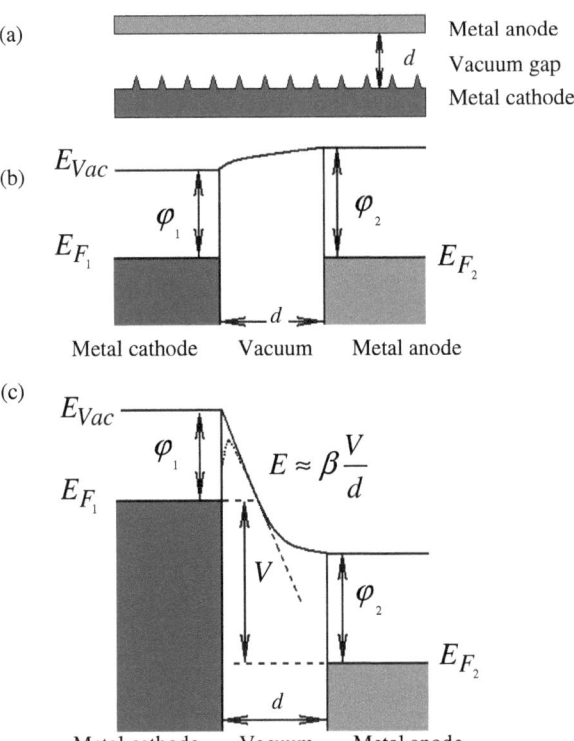

Fig. 2.3: Sharp cone structure of a metal cathode and a planar anode (a), energy band diagram of it at thermal equilibrium (b) and energy band diagram under applied voltage (c).

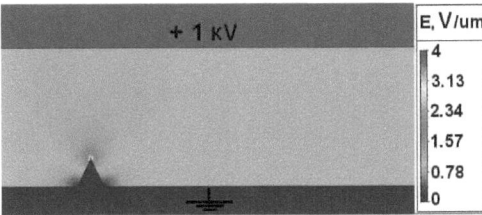

Fig. 2.4: Results of a numerical simulation (ElecNet [Ele]) of electric field distribution between two planar electrodes (1 mm spacing) by presence of a protrusion (left) on the cathode surface showing field enhancement up to 4 V/μm comparing to 1 V/μm on the smooth part of the cathode.

Thus, the FN equation (2.1), which is derived for planar cathode with an assumption that there is uniform electric field in the vacuum gap, cannot be precisely applied. The precise calculation of potential distribution, electric field, and emission current for a sharp microstructure involves numerical calculation of 3-dimensional Poisson equation and Schrödinger equation for electron emission [Gar98, Fur98, Jun98]. An extension of the Fowler-Nordheim treatment was developed by Sommerfeld and Bethe [Som33] to include a factor known as the field enhancement factor, to take into account the fact that the local electric field may differ from the macroscopic value, due to sharp emission tips, which are of utmost importance in FE theory. The emission current for a sharp microstructure can be obtained with a simple modification of the FN equation for a planar metal cathode by replacing the parallel electric field E_0 with local electric field E at the apex of the sharp microstructure as follows:

$$I_{FN} = A \frac{S(\beta E_0)^2}{\varphi\, t^2(y)} \exp\left(-B \frac{\varphi^{3/2} \upsilon(y)}{\beta E_0}\right) \qquad (2.2)$$

$$E = \beta E_0 = \beta \frac{V}{d} \qquad (2.3)$$

where E_0 is the macroscopic electric field and β is defined as the geometrical field enhancement factor, which is the factor of which electric field is increased due the sharp microstructure relative to the planar structure. High β factor and resulting microscopic field lead to much reduced potential barrier width (see Fig. 2.5) and correspondingly enhanced FE due to increased tunneling probability of electrons.

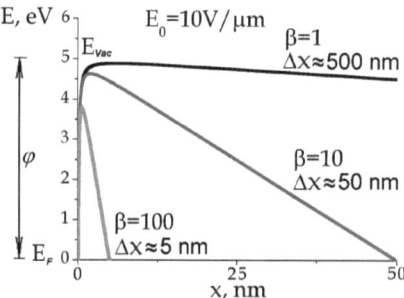

Fig. 2.5: An effect of the field enhancement factor β on the potential barrier in a conductor with work function $\varphi = 5$ eV at macroscopic electric field $E_0 = 10$ V/µm resulting in much reduced barrier width Δx given for electrons of the Fermi level.

The β and S parameters can be determined from experimental data plotted in so-called FN coordinates i.e. $\ln(I/E^2)$ vs. $1/E$ (FN plot). Straight line of the FN plot indicates that FE is the dominant process and slope of this line $K = d(\ln(I/E^2))/d(1/E)$ and ordinate crossing C (or its linear fit) give values of the β and S parameters assuming known work function value:

$$\beta = -\frac{B\varphi^{3/2}}{K} \quad (2.4)$$

$$S = \frac{\varphi \exp(C)}{A\beta^2} \quad (2.5)$$

It is well known that the geometrical field enhancement factor increases with sharpness of the tip and the field at the apex of the tip is inversely proportional to the tip's radius. This simple approximation implies that the emission current for a sharp microstructure is equivalent to the emission current of a planar cathode of the same vacuum gap but the effective electric field is increased by the factor of β. This approximation agrees very well with experimental results because the electric field of a sharp tip is stronger at the apex and reduced rapidly for the region away from the apex and thus it can be assumed that most of emission current arises from electron tunneling within the vicinity of this highest electric field region. In order to estimate the value of the factor β, different field enhancement models have been introduced. A simple field enhancement model [Orv89, Cui01, Uts91] can be applied to microstructures with certain geometry with a smooth surface. The field enhancement factor β for various shapes (Fig. 2.6) of microstructures has been estimated based on electrostatic theory.

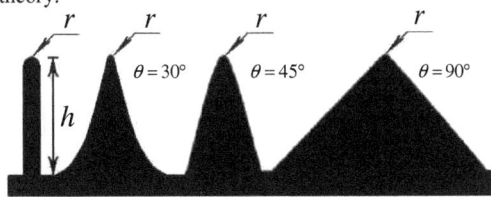

Fig. 2.6: Various shapes of field emitters: rounded whisker (a), sharpened pyramid (b), hemispheroid (c) and pyramid (d)[Orv89].

The electric field at the surface of a sphere (Fig. 2.7(a)) of a rounded whisker can be evaluated using elementary electrostatic theory and expressed in closed form as a function of polar angle θ as

$$E = E_0 \, (h/r + 3\cos\Theta) \tag{2.6}$$

$$E_0 = \frac{V}{d} \tag{2.7}$$

$$\beta = h/r + 3\cos\Theta \tag{2.8}$$

where E_0 is macroscopic field between two planar electrodes, h is the height of the microstructure from the base and r is the radius of the tip. For $h \gg r$, $\beta \approx h/r$. It has been shown that the field at the apex of a rounded whisker shape is approximately equal to that of a floating sphere and is given by $E = E_0 \, (h/r)$. Thus, it has primarily concluded that the round whisker shape is the closest to the "ideal" field emitter. On the contrary, a wide-angle pyramidal shape is not an optimum field emitter even though its thermal and mechanical stability is excellent.

Two-steps or protrusion-on-protrusion field enhancement model is a modified version of the simple field enhancement model developed to account the complicated morphology of emitters. The emitting surface may be described as a number of small protrusions act as tiny tips. The emitting tip with height h_1 and sharpness of radius r_1 is assumed to consist of a number of tiny tips with height h_2 and sharpness of radius r_2 as shown in figure 2.7. The electric field on the blunt tip is equal to

$$E_1 = \frac{h_1}{r_1} E_0 \tag{2.9}$$

and the field at the end of protrusions is equal to

$$E_2 = \frac{h_2}{r_2} E_1 = \frac{h_2}{r_2} \frac{h_1}{r_1} E_0 = \frac{h_2}{r_2} \frac{h_1}{r_1} \frac{V}{d} \tag{2.10}$$

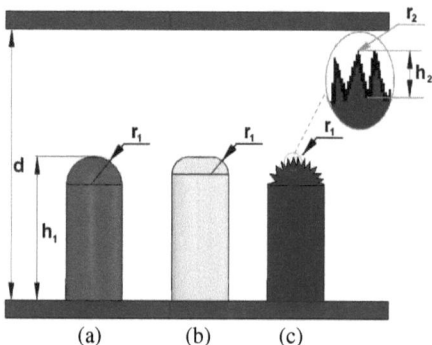

Fig. 2.7: Schematic representations of simple (a, b) and protrusion on protrusion (c) field enhancement models [Giv95].

With this model, it is easy to understand a low electric FE from blocks or columns of densely packed CNT discussed below (see Chapter 4). It indicates dominant emission enhancement by outstanding individual CNT rather than by geometry of the full structures and it makes aspect ratio of the full structures nearly irrelevant for the field enhancement.

Discussing parasitic FE, there are few more models explaining possible mechanisms of the emission from metallic surfaces. Models presented below have their beginning in studies of pre-breakdown phenomena in vacuum-insulated high-voltage electrodes systems [Lat95] and can be transferred to explain parasitic FE from superconducting (SC) niobium cavities required for particle accelerators [Chapter 5]. In the vacuum-insulated high-voltage electrodes systems the current was found to originate, by macroscopic fields of 5-20 MV/m, from non-metallic protrusions on the cathode surface [Lat95]. This was originally explained in terms of field enhancement by a factor of about 200-300, which creates the barrier field of 3 GV/m required for the FN tunneling mechanism.

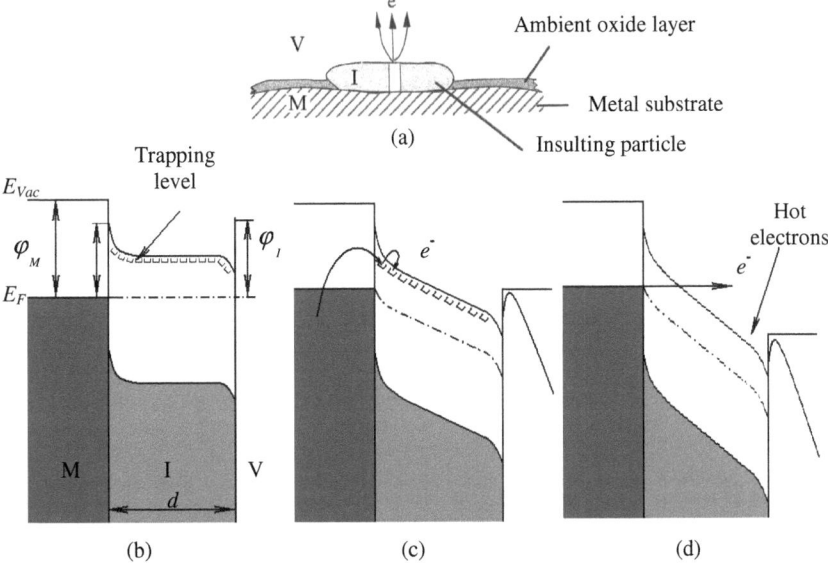

Fig. 2.8: *A schematic illustration of the MIV emission model (a) and a sequence of band diagrams illustrating the FE switch-on process: thermal equilibrium at no field (b), electron injection by hopping at medium field (c), electron tunneling and generation of hot electrons leading to burst of electron emission at switch-on field (d) [Lat95].*

However, other studies of the emission process suggested that it is associated with some form of dielectric surface micro-inclusions. Evaluating these findings, Latham et al. proposed an alternative emission mechanism, based on field-induced electron heating [Lat95, Xu95]. This assumed a 'metal–insulator–vacuum (MIV)' emission regime, associated with field penetration into the surface 'insulator', and the formation of a conducting channel. Figure 2.8 shows a schematic illustration and band diagrams for this model. The penetrating field was assumed to heat electrons, so that (at least to some extent) they could be emitted over the surface potential barrier rather through it by FN tunneling.

Unlike the protrusion field enhancement model described above, field electron emission in the MIV model cannot occur even when the surface potential barrier at the insulator-vacuum interface is narrow enough for conduction electrons to tunnel it, since very few electrons exist behind the barrier. Thus, it is found that such an emission regime needs to be "switched on" in order to give the rise to a steady emission current. Two conditions are usually required for the occurrence of such a switch-on process: namely, a voltage "surge" superimposed on some threshold gap voltage. This switch-on state normally persists even when a sample left for a long period without the application of electric field under UHV conditions at room temperature, i.e. indicating some permanent formation of one or more electron conducting channels in the bulk of the insulator. Switch-on process of electron FE was often observed during investigations of parasitic FE from Nb surfaces what will be presented and discussed in the Chapter 5.

Metal-insulator-metal (MIM) or hot electron emission model proposed that the mechanisms of parasitic FE from metallic surfaces with conducting contaminations involve the creation of electroformed conducting channels via hot electron emission process in metal-insulator-metal (for example, niobium - niobium oxide - conducting particulate) microstructures. This model arises from experimental observation on the FE characteristics of diamond-coated Mo emitters [Xu94] and was also proposed to explain how carbon graphite particle artificially deposited on a Cu electrode could promote low field (< 10 MV/m) 'cold' electron emission [Xu95, Ath85]. The emission characteristic of the diamond-coated Mo emitters has a broad electron energy spectrum (~1 eV), which is larger than of a clean Mo tip of ~0.23 eV with a large spectral shift of ~2-3 eV (the different between center of energy spectral and the Fermi level of the Mo substrate). These emission characteristics are similar to the emission spectra of the metal-insulator-metal (MIM) graphite microstructure, which has been explained by the hot electron emission model [Mou07, Xu86]. It has also been found that such MIM structures are the predominant electron emission sources on heat-treated broad area niobium electrodes [Mah93]. The hot electron emission model for MIM microstructure is illustrated in figure 2.9. In this structure, it is assumed that a particulate placed on an oxidized metallic surface act as an "antenna" by probing the field above the electrode surface, and thereby producing an enhanced field across its contact point with the insulator. If the particulate has height is h and the insulator thickness is d then the externally applied field is enhanced by a factor on the order of h/d in the insulator region, and thereby resulting into an enhanced field across its contact point with the insulator. Since the oxide will block the transportation of carriers from the substrate metal, this field enhancement will eventually lead to a significant voltage drop across the oxide layer sandwiched between the particle and the metal. Thus, in a switch-on process, a conducting channel is preferentially formed in this region. The energy band diagram of the MIM structure is illustrated in figure 2.9(b). Under an applied field condition, electrons will tunnel from the metal substrate into the conduction band of the insulator by a tunnel-hopping process and subsequently accelerated in the channel by the internal field to become "hot electrons" i.e. by the same MIV mechanism described above. At the top metallic layer, the electrons that are heated internally will undergo a coherent scattering process [Lat95], which has been explained by electron diffraction model [Sim67], so that they can be emitted into vacuum without loosing the kinetic energy gained from the field passing through the insulating layer.

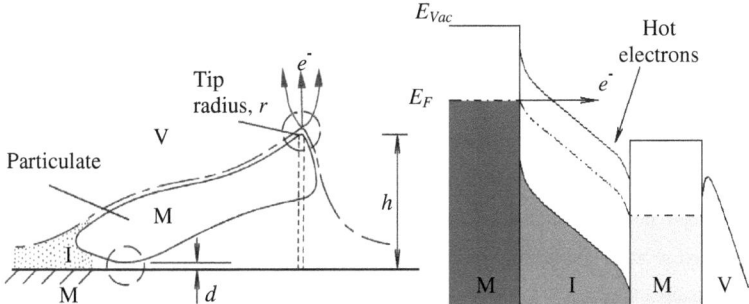

Fig. 2.9: A schematic representation (left, [Mou07]) and a band diagram (right) of the MIM structure explaining low-macroscopic-FE from a conducting particulate microstructure on a broad-area planar electrode.

If an atom or molecule is adsorbed on the surface of the emitter it modifies the potential at the emitter-vacuum interface, and can produce surface states through which resonant tunneling can occur [Duk67, Gad93]. This process is presented in figure 2.10. Resonant tunneling can increase the locally emitted current by orders of magnitude while maintaining a narrow energy spectrum. The current through a single adsorbate can extend well into the microampere regime, depending on the adsorbates binding energy. With this current level, the narrow spectrum (~0.5 eV) and spatial localization of the enhancement (~1 Å) results in beam brightness near the quantum limit [Jar10]. Because of such complex structure of energy band diagrams of emitters, in some cases an alternative to the FN equation physical basis has to be provided for correct analysis of the emission mechanism. Use of FN analysis in these cases leads to unreasonable values, for example of the field enhancement factor β or effective emitting area s.

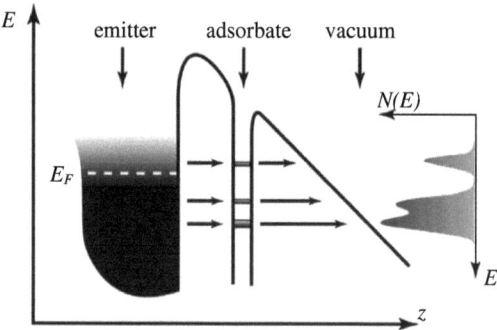

Fig. 2.10: Energy diagram of the emitter-vacuum interface in the presence of an adsorbate and the resulting FE energy distribution [Jar10].

2.2. Semiconductor and band structure effect

According to the FN model, FE from metal is determined mainly by the barrier at the solid-vacuum boundary. The work function of the metal and the electric field at the emitter surface determine the potential barrier. FE from semiconductors, relative to metals, is a much more complicated process due to the low carrier concentration in the emitter bulk. The low carrier concentrations allow for field penetration into the semiconductor, causing band bending and nonlinearity of the current-voltage characteristics in FN coordinates. Under certain conditions, these features can make the FE current very thermal- and photosensitive. The capture of free carriers by traps can cause a deviation from the electrical quasi-neutrality in the emitter bulk and lead to modifications of the field distribution at the surface [Fur05]. The interest in FE from semiconductors is stimulated by many reasons. Distinct from metals, a semiconductor offers numerous ways of varying the characteristics of the emission process by control of carrier concentration in the emitter bulk, thus making unique electron devices possible. Application of semiconductors in vacuum microelectronics is aided by the fact that for some semiconductor materials, such as Si, the basic technology of fabricating complex structures has been developed. Analysis of parasitic FE, for example, from Cs_2Te

photocathodes (see Chapter 3.3) requires also the clear understanding of the FE processes from semiconductors. Another important aspect is that the CNT, which are actually the most prospective material for cold cathode applications, can exhibit both metallic and semiconducting properties [Ado08] depending on their chirality and number of layers. Single-walled nanotubes (SWNT) are an important variety of CNT because they exhibit electric properties that are not shared by the multi-walled carbon nanotube (MWNT) variants. In particular, their band gap can vary from zero to about 2 eV and their electrical conductivity can show metallic or semiconducting behavior, whereas MWNT are zero-gap metals [Ado08, Bus10]. Most of works on CNT cathodes reported so far utilized CNT deposited on semiconductor (mostly Si) substrates or integrated in semiconductor multilayer structures using well-established semiconductor technology. Therefore, peculiarities of FE from semiconductors including band structure and carrier concentration effects should be taken into account analyzing FE results and will be shortly discussed in this chapter.

The relationship between the FE current and electric field (or voltage) for semiconductors has been the subject of a number of investigations [Mod84, Fis66]. It was found, in contrast to metals, that in semiconductors with low carrier concentrations in the conduction band (high-resistivity p- and n-types) the current-voltage dependence in FN coordinates is nonlinear [Yos10, Lu06]. Some evidence of nonlinear current-voltage characteristics appear in early studies by Apker and Taft who investigated CdS [Apk52]. The first comprehensive study of the nonlinear current-voltage characteristics was carried out by Sokol'skaya and Shcherbakov [Shc62] and showed that the current-voltage curve can be divided into three regions (Fig. 2.11): region I - linear current-voltage dependence; region II - slow current variation ("saturation" region [Dea00]) - in this region a strong photo- and thermal sensitivity of the FE current has been observed [Liu06b, Yos01]; region III - rapid rise of the current with voltage (in some cases more rapid than in region I).

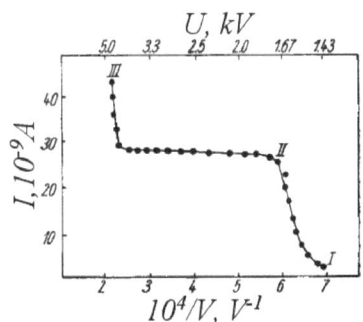

Fig. 2.11: An example of nonlinear current-voltage characteristics of semiconductors (p-type Si) [Fur05].

In intrinsic semiconductors at room temperature, the Fermi level is located in the forbidden band gap between valence and conduction bands and the conduction band is rarely occupied. Therefore, electrons can be emitter only from the valence band. When $kT \ll E_G$, where T is the temperature and E_G is the band gap of the semiconductor, the conduction band is empty. In that condition, FE occurs by tunneling from the valence band to the vacuum if the applied field is high enough. Since each emitted electron leaves a positive hole in the surface of the material, the emission current is balancing by the conduction provided by the holes [Gom93].

Considering extrinsic semiconductors, it is demonstrated [Fur05] that in p-type semiconductors, near the crystal boundary, a weakly conducting region is formed, depleted of mobile carriers which exerts a limiting influence on the current and results in a saturation region appearing in the current-voltage curve. The saturation current value depends, essentially, on the generation rate of mobile carriers. In n-type semiconductors, this kind of region is not forming, and the phenomenon of saturation arises due to the limited generation rate throughout the whole depth of the sample. The energy diagram of the near-surface region of the semiconductor in the electric field is given in Fig. 2.12. Here $\theta(x)$ is a function characterizing the position of the bottom of the conduction band relative to the Fermi level in

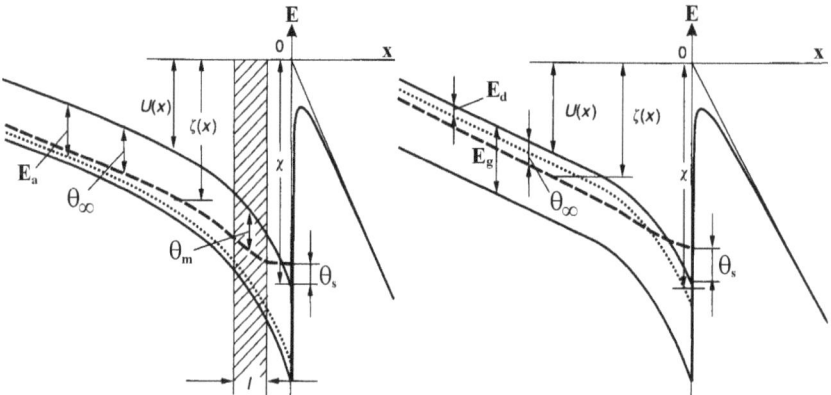

Fig. 2.12: Energy band diagram [Fur05] of a p-type (a) and n-type (b) semiconductor in a high electric field.

any point of the crystal, θ_s the value of this function on the surface, θ_∞ in the bulk of the sample, $U(x)$ and $\zeta(x)$ are the energies at the bottom of the conduction band and of the electrochemical potential level, respectively, which are referred to an electron at rest at infinity, χ is the magnitude of the electron affinity, E_a and E_d are the energies of the acceptor and donor levels, referred to the bottom of the conduction band.

Many effects like charge redistribution and band bending due to field penetration [Che04, Koe06], influence of space charge effect on the emitted current [Jen09], generation-recombination processes [Sno09], an increase of the voltage drop on the emitter due to the increase of the current flowing through the sample [Fur05], temperature dependence of FE current [Zhu09], concentration of free electrons and the existence of capture centers, including deep traps, in the semiconductor [Fur05] etc. make the matter very complicated, but have to be considered for clear theoretical understanding of FE from semiconducting materials.

2.3 Cold cathodes with mutual shielding

For cold cathodes with closely packed arrays of nanostructures, in which the separation between individual nanostructures d is much less than their height h, it was found that mutual electrostatic shielding can dramatically affect the FE performance of the cathodes [Grö00, Nil00, Bon01, Mcc07, Suh02, Mil04]. The equipotential lines over the emitters change due to neighboring emitters in the manner shown in Fig. 2.13, and thus the corresponding field enhancement factor β is also affected strongly by the variation in inter-emitter distances. When nanostructures are far apart, the field enhancement is strong, but the total number of emitters per unit area is low, which reduces the emitted current. When nanotubes are close together, they shield each other, reducing the field enhancement factor. Several numerical simulations and experimental results showed that at $d/h \approx 2$, an optimal spacing is reached where the field is only minimally reduced by the neighboring nanotubes, and their numbers per unit area remain high [Mil04, Dio08, Kim06a]. Therefore, to produce the most effective FE cathodes, we must create aligned, patterned arrays of nanostructures at controlled spacing.

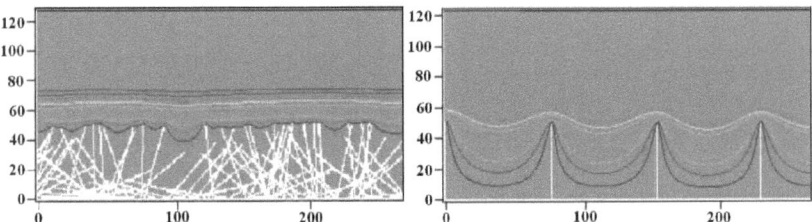

Fig. 2.13: Simulations of the equipotential lines of the electrostatic field showing shielding effect of dense CNT (left) and enhanced electric field on spaced apart CNT with minimized shielding [Mil04].

2.4. Electro-thermal properties of nanostructures

MNW and CNT are regarded as key components of future nanoscale devices. Among other important factors, their electro-thermal properties are crucial for a reliable performance of nanostructure-based modules. Recent experimental and theoretical results demonstrate that so-called Rayleigh instability causes the break-up of a cylindrical MNW into a linear row of nanospheres (Fig. 2.14). It happened at temperatures much below the bulk melting point, but where atomic movements by diffusion become significant [Toi04, Kar06, Mül02, Kar07].

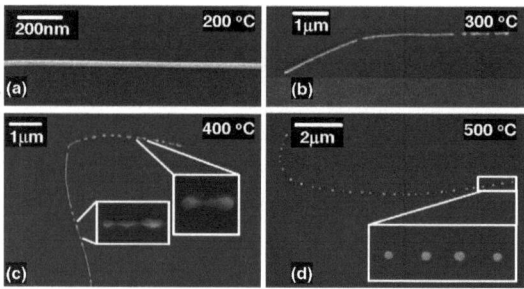

Fig. 2.14: HRSEM images of 25 nm gold wires after annealing at different temperatures for 30 min [Kar07].

This poses a serious obstacle to the sustained reliability of components basing on nanowires (NW). The driving force for such changes in the microstructure is the reduction of the surface energy of the system. In any rod-like morphology, the surface energy can be reduced by the spheroidization. The morphological stability of continuous and high aspect ratio structures is a problem with long-standing interest in material science and technology [Kar07].

When the size of an object becomes comparable to the electron mean free path, its

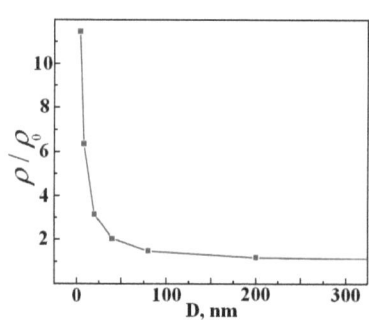

Fig. 2.15: The resistivity ratio for thin Au wires ρ and bulk material ρ_0 plotted versus diameter of the wires. The curve is calculated using Dingle's model [Kar07].

electrical transport properties are influenced rather strongly by the electron scattering (elastic or inelastic) from the surface and internal grain boundaries. This gives rise to the so-called finite-size effects. The electrical resistivity of a wire increases once its size becomes of the order of the electron mean free path as shown in Fig. 2.15. The curve is calculated assuming the electron mean free pass of 40 nm for bulk gold. The graph reveals already the drastic increase of the resistivity of gold with decreasing wire diameters by taking into account only electron scattering from the surface [Kar07]. From Wiedemann-Franz law it is known that for metals there is a linear correspondence between thermal and electrical conductivity. Therefore, corresponding decrease of thermal conductivity of wires is expected also.

The electron transport properties of a SWNT (either a metal or a doped semiconductor) deviate significantly from Ohm's law because the charge carriers can flow over a large distance (~1 µm at room temperature), without suffering from inelastic scattering by phonons and other excitations. Even in the absence of any scattering, the resistance of a nanotube is not zero. The reason is that the nanotube offers only a few propagating states in comparison with macroscopic contact electrodes that have many states. At small bias, the conductance of the nanotube is controlled by the probability that an electron entering the nanotube from one electrode with Fermi energy can be transmitted to any state with the same energy in the other electrode. In the case of a perfect metallic SWNT at zero temperature, the electrical resistance is 6.45 kΩ. It is length independent (ballistic transport) and quantized due to the finite number of conducting channels. In practice, the resistance of a metallic SWNT can be significantly larger when there are poor coupling contacts with the external electrodes. In addition, bending, squashing, or twisting a nanotube may affect its conductivity, sometimes inducing metal-semiconductor transformations [Bus10].

The electron transport regime in MWNT is not established as well as for SWNT. Experiment shows that they have a wide spectrum of transport properties. Some MWNT are metallic or semi-metallic, others are clearly nonmetallic. It is generally admitted that electron transport in a metallic MWNT at room temperature is not ballistic but instead, diffusive [Bac00]. However, evidences also exist for ballistic transport taking place in the outermost shell of an individual nanotube [Fra98]. The room-temperature resistivity of annealed, large MWNT is 2×10^{-6} Ωm and their resistance per unit length is typically 30 kΩ/µm. They can carry a very large current density (10^{11} A/m^2) without failure. A single nanotube can emit a stable current of a few microamperes [Dea01, Sem02]. As well as mechanical stiffness and thermal stability, CNT possess a high thermal conductivity along the tube, surpassing diamond's value ~ 3000 W/m·K (compare to copper 385 W/m·K) at room temperature (Fig. 2.16), due to the strong carbon-carbon chemical bonding [Ber00]. But SWNT is a good insulator laterally to the tube axis showing 1.52 W/m·K thermal conductivity at the room temperature. Due to high thermal conductivity, the heat transport in CNT has attracted much attention from both theoretical and experimental points of view. Heat transport in the case of SWNT as well as MWNT is predominantly carried by phonons [Jor08]. In contrast to SWNT, MWNT with a large diameter cannot be regarded as simple one-dimensional phonon systems, and exhibit complex heat-transport phenomena because of the van der Waals interaction

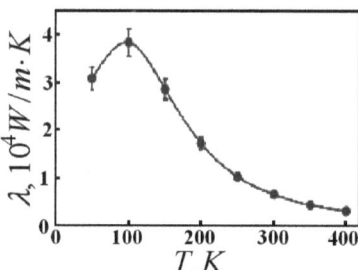

Fig. 2.16: The thermal conductivity of CNT as function of temperature [Ber00].

between the tube walls. At low temperatures, phonon conduction is ballistic through the entire body of the SWNT, leading to universal quantization in the thermal conductance. As the temperature is increased, the length of the phonon mean-free path becomes comparable to that of SWNT, and phonon conduction ceases to be ballistic. The heat transport behavior changes from quasi-ballistic to diffusive at temperatures above the room temperature. As such, a marked reduction is observed in the thermal conductivity due to defects and contacts to the CNT. Thermal conductivity of the CNT seems to change also depending on current and vacancy concentration [Jor08].

Comparing the electro-thermal properties of CNT and MNW, the CNT look more promising for cold cathode applications because of higher current carrying capability as well as thermal and mechanical stability.

3. Measurement techniques

The FE measurements were executed with two instruments, i.e. the FESM for spatially resolved measurements and the IMLS for applicational issues. Detailed description of the instruments can be found in following works [Hab98, Göh00, Lys05, Lys06, Mah95, Pup96]. In this chapter only a short description including some changes i.e. improvements of the systems performed during the current research will be mentioned. The new SAFEM will be presented in more details because it was a significant part of the thesis work.

Correlations between FE results and surface morphology/defects were made by means of optical profilometer (OP), atomic force microscope (AFM) and scanning electron microscope (SEM) including energy dispersive x-ray analysis (EDX). Short description of the instruments will be given.

3.1. FE scanning microscope (FESM)

The FESM (Fig. 3.1) is an advanced microscope for investigation of FE site distributions on "flat" cathodes of up to 25 x 25 mm². The cathode is xy-surface tilt-corrected

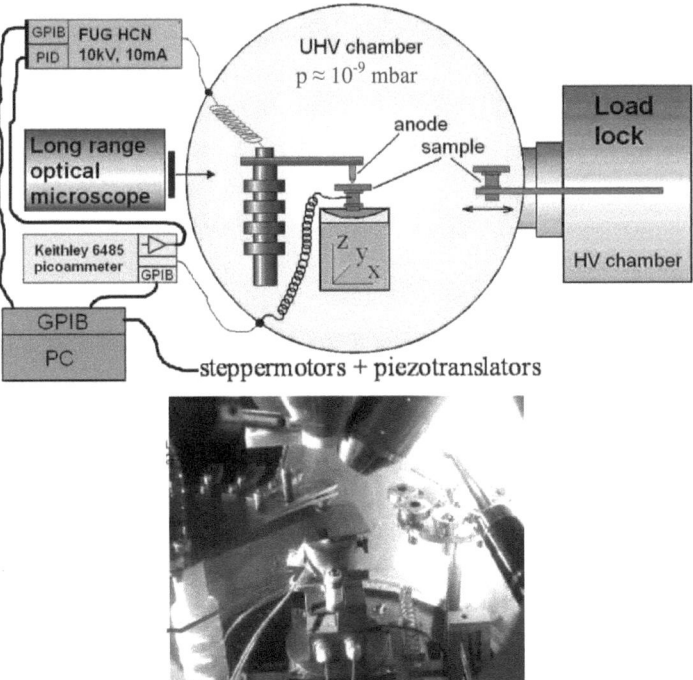

Fig. 3.1: Schematic illustration (left) and inner view (right) of the FESM.

with respect to the anode to achieve a constant gap Δz within ±5 μm for the full scan area. The measurements are done in ultrahigh vacuum (UHV) of 10^{-9} mbar by means of non-destructive regulated voltage scans V(x, y) at a fixed FE current (from 1 nA to 10 mA). It employs a PID-regulated power supply FUG HCN100M-10000 [Fug] controlled by the FE current measured by a digital picoammeter Keithley 6485 or an analog electrometer Keithley 610C [Key] as described elsewhere [Lys05, Lys06]. The power supply provides up to 10 mA current and 10 kV voltage, which corresponds to 500 MV/m macroscopic field at 20 μm gap.

By in-situ exchange between needle-anodes and flattened conical anodes, the alignment, homogeneity, and efficiency of FE from selected cathode areas can be verified with the required resolution, and the integral current of arbitrarily chosen individual emitters or patches can be measured. The needle anodes have a tip apex radius from $R_a \geq 1$ μm, while the flattened conical anodes are up to 300 μm diameter. The anodes are mechanically prepared and finally smoothed or, in case of the needle-anodes, sharpened by means of electrochemical etching in 10% water solution of NaOH and consist of tungsten-tantalum alloy 90:10. The local FE measurements give values of the onset field E_{on}, an effective field enhancement factor β_{eff} and the maximum current capability I_{max}. The macroscopic electric field E is calibrated for each emitter as the linear slope of the PID-regulated V(z) dependence for 1 nA current. Moreover, the real distance d between the tip anode and the relevant emitter is determined by the linear extrapolation of each V(z) curve to zero voltage (Fig. 3.2, [Nav09a]). For d values close to or smaller than the mean height h of a nanostructure, the effective field enhancement factor β_{eff} is not only given by the intrinsic field enhancement of the nanostructure (see Chapter 2), but is also influenced by the geometric field enhancement β_{an} provided by the nearby anode as shown in the Fig. 3.3. The reduction of β_{eff} with d was confirmed experimentally for a single CNT [Bon02] and can be estimated based on a two-region FE model of *Zhong et al.* [Zho02] as

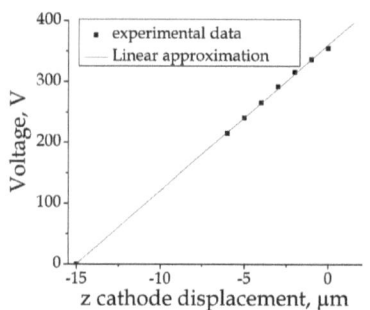

Fig. 3.2: Determination of real distance d between the anode and e relevant emitter.

$$1/\beta_{eff} \sim 1/\beta_{ns} + 1/\beta_{an} = r_k/h + r_k/d \qquad (3.1)$$

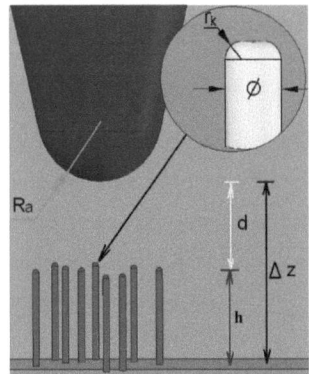

Fig. 3.3: Schematic of the measurement geometry of nanostructures with a needle anode defining all important parameters.

This model is not exact for patches of nanostructures yet because of influences of mutual shielding and field reduction α due to the anodes geometry [Lys06] on β_{eff} have to be taken into account, too. The decrease of the field enhancement factor at decreased distance d has been well confirmed experimentally by measurements on randomly distributed Au-NW [Dan08]. Therefore, the effect is very important for correct design of triode structures of real applications.

During the thesis work, several repairs of the microscope have been done including replacement of a high voltage insulator, replacement of ball bearings and some parts of sliding stages, installation of a new high voltage power supply etc. The insulator part is made from a machinable and UHV compatible Macor® ceramic and was lathed in a machine shop of the University of Wuppertal.

An attrition of the ball bearings caused a jamming or springing of the sliding stages leading to the appearance of artifacts on the voltage maps (see for example Fig. 4.8(b)) or even to touching between electrodes causing shorts and cathode destruction. The balls are made from the hard metal alloy (91-92 % tungsten carbide + 8-9% nickel), have 3.17 mm diameter (kept with a very high precision G10 DIN5401) and were replaced completely in all three stages. To prevent fast attrition of the balls and sliding guides, an UHV compatible lubricant "Ultratherm S3002" [Lub] was used. The lubricant can be heated up to +280 °C and can be used in vacuum down to 10^{-12} mbar. Every repair of inner parts and exposure them to the ambient air were followed by several days of exhausting process including careful heating of the full vacuum chamber up to maximum +150 °C to reach the required vacuum level.

In order to reach higher field levels, required especially for parasitic FE investigations, the 5 kV power supply was replaced by the 10 kV one mentioned above. The LabView programs of the FESM were correspondingly changed because of more modern version of the GPIB interface and commands of the 10 kV power supply. Additionally few new very useful functions have been added to the programs. It includes automatic rise and fall of the voltage up to a maximum current/voltage with specified speed, registration of current and voltage as time-reference data required for long-term processing of individual emitters, imaging of I-V, I-E, I-t, V-t plots as well as voltage or field maps comparing to only I-V and field maps of previous version, etc.

One more renew of the FESM, which has been started within the time of the thesis work and should be finished in the nearest future, is repair and reuse of an in-situ SEM in the FESM. The SEM should significantly increase a positioning accuracy of the anode over a sample, help for better re-identification of emitters, and give possibility to make some investigations of surfaces of samples. The SEM has about 1 µm resolution. The resolution is limited by the finite electron gun distance of about 3 cm but it should be enough to improve the positioning accuracy, which actually is about 100 µm (see Chapter 5.3). A tungsten filament of the electron source of the SEM is broken and will be replaced by a more robust and advanced one from LaB_6.

One of the difficulties of current-voltage (I-V) measurements of the FESM is a possible breakdown during measurements of low cathode currents, i.e. at very sensitive ranges of the ammeter, at presence of a high voltage on the anode. Several times the electrical breakdowns led to destructions of input circuits of the picoammeter. Industrial repair of it is very costly but, fortunately, several times it was successfully repaired by myself. To avoid further destruction of the picoammeter, a protection circuit (Fig. 3.4) was built and successfully used during the local I-V measurements. The protection circuit limits the voltage applied to the input of the ammeter to around ±0.7 V, which is equal to a voltage drop on the open diode p-n-junctions. Maximum continuous input signal of the Keithley 6485 picoammeter is 220 V peak, dc to 60 Hz sine wave. To limit the current through the diodes (IN3595), a current limiting circuit of the FUG power supply and additionally a resistor R = 12 kΩ were used. To avoid any influences on the obtained results, the diodes should have an extremely low leakage current (~ pA) and should be in a light-tight enclosure to prevent light induced leakages. Such protection of the ammeters in the scanning mode, however, is not possible, as it increases time delay by introducing additional resistance and capacity and decrease the dynamic range of the ammeter. Therefore, the voltage scans were always received without any protection. It imposes constraints on use of the old analog ammeter Keithley 610C, because industrially it is not supported any longer and no spare parts are available. In case of a failure, it has to be replaced by a modern one. Some voltage scans were done already with the digital Keithley 6485. Advantages of the 610C ammeter are

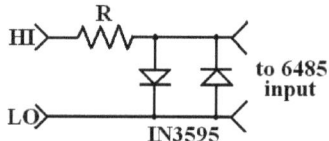

Fig. 3.4: Overload protection circuit of Keithley 6485 picoammeter.

robust input circuits and possibility of direct control of the FUG power supply by an analog output signal of it. The analog output signal of the 6485 one, contrariwise, has to be additionally converted (divided by -2.5). Converters used so far, however, introduce additional noise and error. Therefore, two things have to be solved in the nearest future: a protection of the input circuits and a more reliable control of FUG power supply by a modern ammeter.

3.2. Integral measurement system with luminescent screen (IMLS)

The IMLS is used for determination of the integral current of the cathodes in dc and pulsed modes up to 5kV voltage (17 V/µm field at 300 µm gap) and 50mA current. The signal registration of the system was significantly improved comparing to the previous version described in [Lys05]. Serial resistors were used for the cathode protection against discharges (Fig. 3.5(a)) and fast current readout (Fig. 3.5(b)). A resistive voltage divider (Fig. 3.5(c)) is used for the high voltage readout. Registration of current and voltage signals on a PC (Fig. 3.5(f)) is done via an analog to digital (ADC) converter (Keithley KPCI-3102) after a unity gain amplifier (Fig. 3.5(e)). The ammeter (Fig. 3.5(d)) can be used for the current calibration and dc measurements (via GPIB interface). For the pulsed measurements, however, it is not suitable because of its limited speed. The IMLS was correspondingly equipped with a data acquisition system based on a self-developed software using LabView

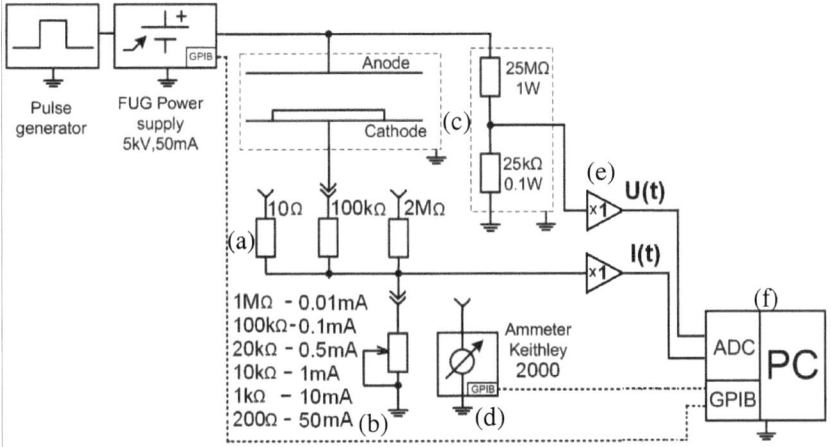

Fig. 3.5: Block diagram of the electric circuit of the IMLS in diode configuration.

programming package. It allows registration of full signals of the current and voltage in the dc and pulsed modes as well as their time dependence during the long-term tests. The system can be switched from the diode to a triode configuration as shown in the Fig. 3.6. However, the triode configuration has to be still improved and it is planned to be done in future works for tests of cathodes with matched gates for the real applications.

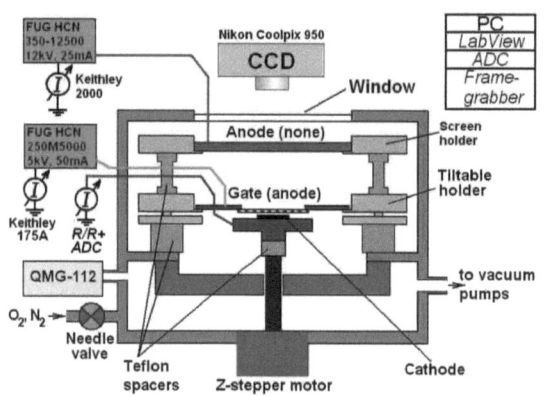

Fig. 3.6: Triode (diode) configurations of the IMLS.

The FE current distribution from the cathodes shown by the luminescent screen can be stored with a CCD camera Nikon Coolpix 950 with around 20 µm lateral resolution. The image signal from the camera is transmitted to a PC via a frame grabber Matrox Meteor II [Mat] and processed with image analyzing software AnalySIS FIVE® [Ana]. It is suggested to replace the camera by an up-to-date one with a much better focusing and a LabView compatible interface. There is an idea to incorporate the acquisition of images into the LabView program and store the images as time- and I-V-referenced data for a faster processing and a smooth correlation between them.

Finally, the long-term performance of the cathodes can be tested in the dc mode at a base pressure of about 10^{-7} mbar as well as under enhanced nitrogen or oxygen pressure levels up to 10^{-5} mbar provided by filtered high-purity gases (99.995 %) through a needle valve for up to 5 hours. A quadrupole mass spectrometer (QMG-112) served both for partial pressure measurements and control of a luminescence layer evaporation with a sensitivity limit of 5×10^{-11} mbar (e.g. for S).

3.3. Novel scanning anode FE microscope (SAFEM)

One major issue of operating laser driven rf guns with high gradients and high duty cycles as electron sources for free-electron lasers like FLASH [Fla] or the future European XFEL [Xfe] is dark current emitted from the rf cavity and the photocathode [Pho]. Imperfect photocathode regions with enhanced field emission (EFE) and their contact area to the rf cavity are considered as main dark current sources at typical macroscopic electric fields of about 40-60 MV/m. In order to ensure low FE photocathodes, investigate dark current sources in details, and improve handling techniques of the cathode, the novel ultra high vacuum (UHV) microscope SAFEM [Nav09c] has been built within this doctoral project. The development was done at Bergische University of Wuppertal on demand and in cooperation with the research center for particle physics German Electron Synchrotron DESY [Des]. It will be a part of the systematic quality control of freshly prepared photocathodes at DESY. Design, control software, and assembly of the SAFEM are actually completed and the SAFEM is under vacuum tests, final tuning, and commissioning steps. First real test of the photocathodes are planned to be done in the nearest future.

3.3.1. Actual status of dark current investigations

In photocathode rf guns, the dark current is defined as "unwanted electrons generated in the absence of the driver-laser pulse." Due to the exponential increase of the FE current with the local electric field according to the FN relation, the most sensitive regions for the dark current can be estimated from calculation of the field distribution in the cavity. MICROWAVE STUDIO [Mws] 3D calculations (Fig. 3.7) of the field distribution in the gun cavity and a part of the coaxial rf input coupler show presence of strong surface electric field at the cathode area (1) and the irises (2).

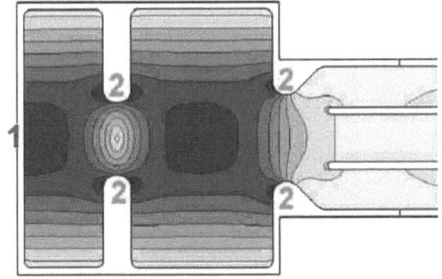

Fig. 3.7: Contour map of the electric field amplitude in the cavity and a part of the coaxial coupler (dark gray correspond to a maximal and light gray to a minimal amplitude) [Jan05].

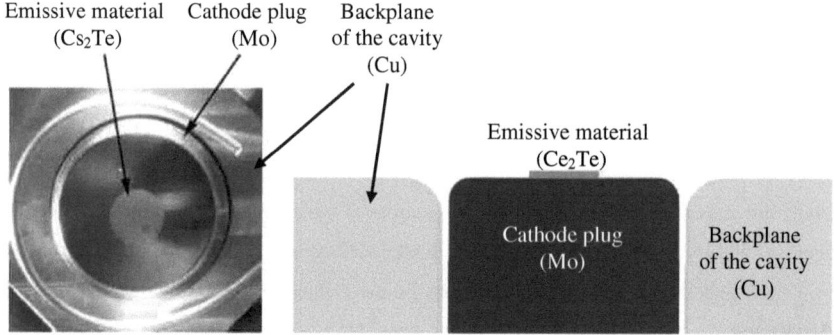

Fig. 3.8: (left) An image of the front surface of the photocathode showing a thin film of Cs_2Te emissive material on a Mo plug, which is fixed in a back plane of the Cu cavity of an electron gun. The circular wire is used only for measurements of quantum efficiency. (right) A schematic cross section of the cathode and the back plane [Jan05].

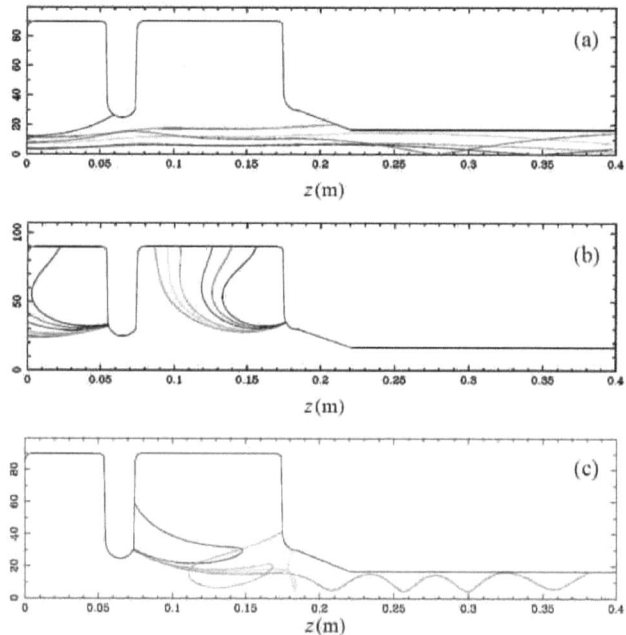

Fig. 3.9: Example trajectories of the electrons emitted from high field strength regions: the cathode area (a) and the first and the second irises of the half (b) and the full cell (c). A gun gradient of 42 MV/m and a main solenoid current of 300 A have been used for the simulations [Jan05].

ASTRA [Flö] simulations of the beam dynamics of the dark current from the regions with high field-strength show that electrons starting at the area of the photocathode [Ser00] (Fig. 3.8) can be accelerated downstream, while field emitted electrons from other sources, like the iris or the entrance to the coupler, cannot leave the coupler (Fig. 3.9). However, they are able to locally heat up the cavity surface, may create secondary electrons, and lead to quenches of superconducting cavities [Pad09]. Therefore, carefully conditioning and EFE investigations of the cavity surface are necessary [Nav09d, Nav10f, Kne95, Dan07a] and will be described in the following chapter. The downstream-accelerated electrons are lost at various places along the beam line and part of them even reaches the undulator [Duk09].

When the dark current is lost, electromagnetic radiation and neutrons are created and may damage diagnostic components and electronic devices installed close to the beam line [Jan05]. The sources of FE current from the cathode area, however, is not studied carefully yet. The FLASH and the European XFEL demand high accelerating gradients and long rf pulses at the gun. Therefore, the amount of the dark current can be comparable to the electron beam current or even higher [Jan05, Jan08]. For imaging of the dark current and the electron beam coming out of the gun, a Ce:YAG luminescent screen [Lip04] can be used. For illustration in the figure 3.10 the Ce:YAG image of the dark current together with the photoelectron beam (central spot) obtained just after the rf-gun exit of FLASH is shown. The rf forward power at the gun was P_{for} = 3.2 MW and the main solenoid current 298 A. The figure 3.11 illustrates only the dark current for the same operational conditions. A strong emitter is visible and the location is assumed to be at the edge of the cathode plug or the rf contact spring area. In addition, a dark current spot exactly at the beam position is present.

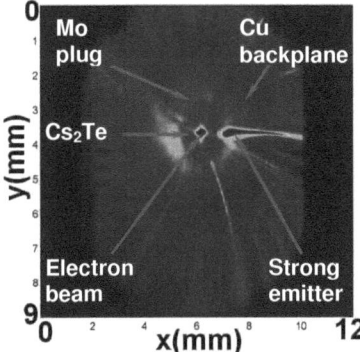

Fig. 3.10: Ce:YAG screen image of the electron beam and the dark current [Sch08].

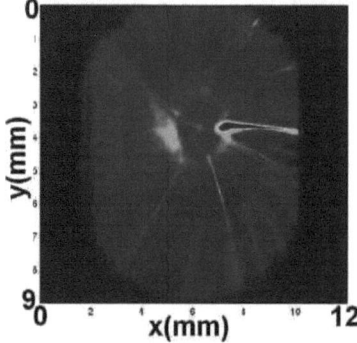

Fig. 3.11: Ce:YAG screen image of the dark current only [Sch08].

3.3.2. Construction of the SAFEM

The SAFEM is developed to be a part of a complex set-up consisting of the preparation system for Cs_2Te photocathodes, the systematic quality control, and the transport system to deliver the photocathodes to the gun (Fig. 3.12). Cathode movement inside the system is based on a transfer system compatible to those one used at DESY photo-injectors.

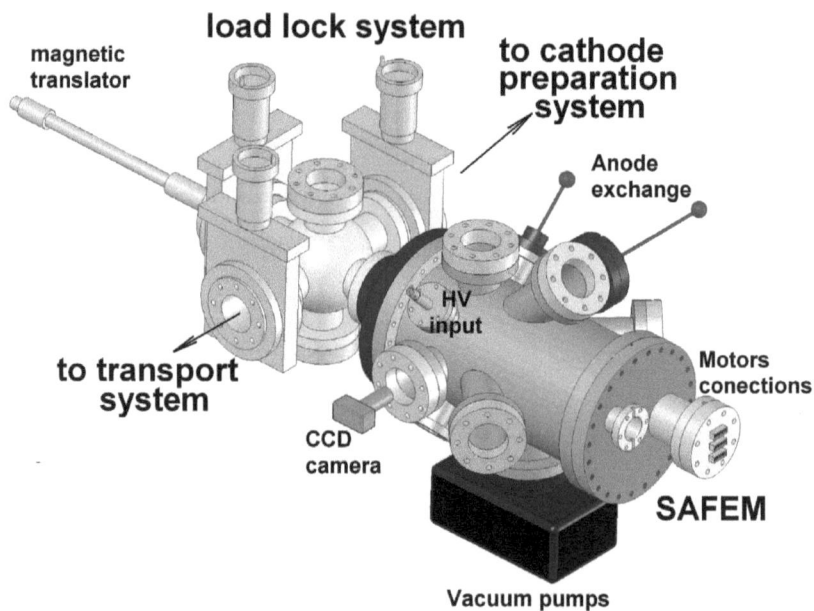

Fig. 3.12: Overview of the SAFEM with a part of the UHV transport system.

Besides FE studies of the dark current from the active photocathode regions, one additional aim of the SAFEM is investigation of handling of the cathodes inside of the rf guns. There is an assumption that scratches on the cathode surface can be produced during installation/removal of the cathodes and lead also to the dark current emission. Therefore, a copy of the Cu gun back plane as a cathode holder inside the SAFEM has been used, including the rf contact spring (Fig. 3.13). The gun back plane is replaced by a Cu plate and original gun springs from CuBe alloy with a silver coating are used. In order to simplify the cathode transfer system and the scanning procedure, as well as to investigate the cathode handling, the design, where an anode of the microscope is scanned over the fixed cathode was chosen. The name "Scanning Anode" was exactly chosen to underline this construction.

Main requirements to the SAFEM are:

- 10^{-10} mbar working pressure,
- 25 x 25 mm^2 scanning range,
- at least 200 MV/m achievable dc electric field strength,
- the electrode gap control to avoid scratches and tilt.

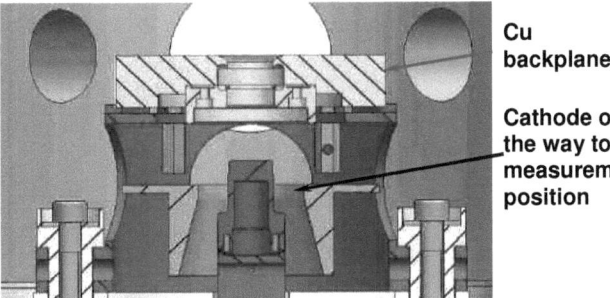

Fig. 3.13: Scheme of the cathode holder inside the SAFEM with Cu back plane.

Such high requirements to the vacuum level are related to the properties of the Cs$_2$Te emissive film of the photocathodes, being very sensitive to vacuum conditions.

The three dimensional scanning system has been constructed on the basis of three UHV compatible sliding stages MTS-65 from Micos [Mic]. Control of the scanning system is done through a special "SMC corvus" 3D controller and a computer via the GPIB interface (IEEE 488). Key parameters for the sliding stages are:

- 25 mm working moving range,

- 0.1 µm resolution,

- 10^{-10} mbar working pressure.

As a voltage source the 10 kV (± 0.1 V, 10 mA) power supply from FUG [Fug] (HCN 100M-10000) with fast PID regulation is used. Using a 20 kV electrical feedthrough on DN40CF flange and CuBe springs, the voltage is supplied to the moving anode. The anode holder and 3D scanning system is electrically isolated using a massive insulator of ribbed form for reduced surface leakage current (Fig. 3.14, 3.15). The insulator is made from the UHV compatible and machinable glass-ceramic Macor®.

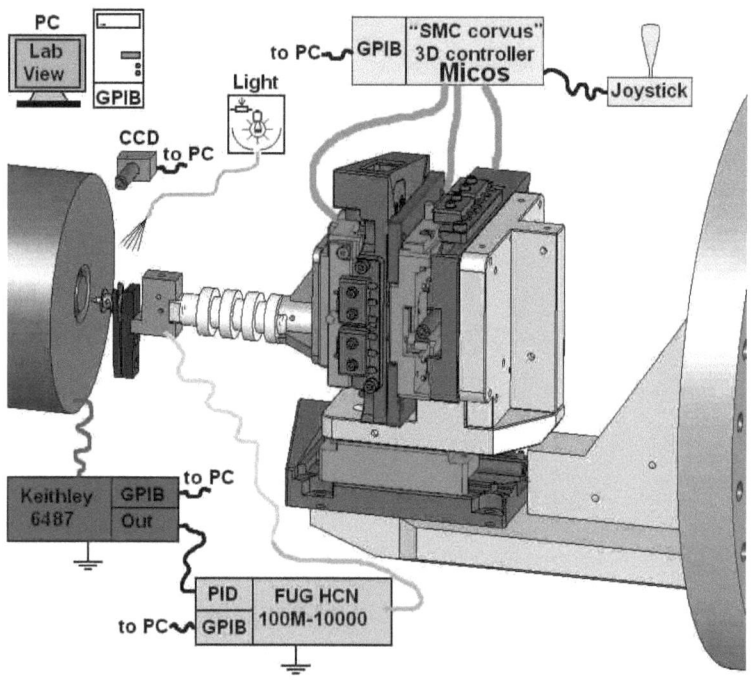

Fig. 3.14: Scheme of the SAFEM including electrical connections.

The FE current is measured between ground potential and the photocathode back plane using the picoammeter Keithley 6487 [Kei]. The copper back plane is electrically connected to the cathode via the silver coated CuBe spring mentioned above but electrically insulated from the transfer system and the vacuum chamber. The electrical connection to the cathode back plane is done via a low-noise electrical UHV feedthrough.

In order to control the electrode gap a CCD camera in combination with a macro zoom lens and a bright light source for illumination are used. The camera and the light source are placed outside the chamber and focused on the electrode gap through view ports.

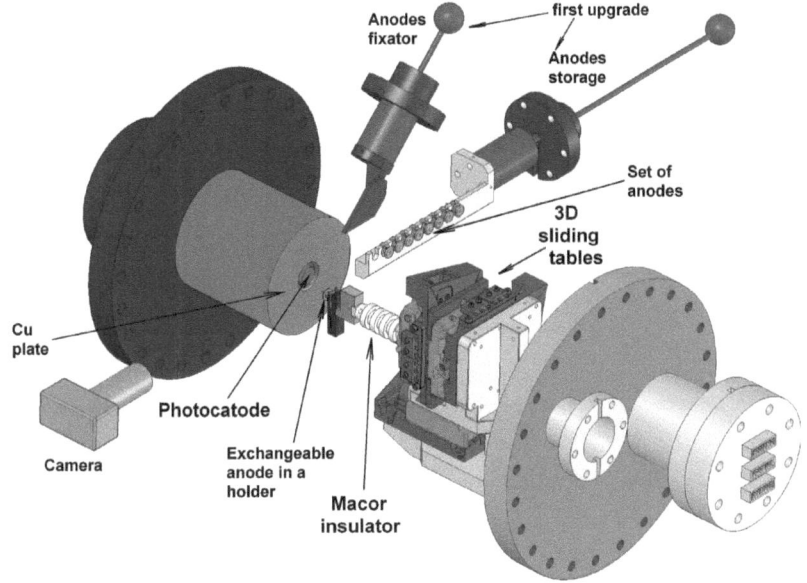

Fig.3.15: Inner view of the SAFEM, already including the future anode exchange option.

Control of the SAFEM in measurement mode is fully automated using a software package (see Appendix A) developed by means of visual programming language LabView. Communication of all instruments is done via the GPIB interface.

The required vacuum level is achieved by using an UHV chamber, only UHV compatible materials, suitable material combinations and parts for all inner constructions of the microscope, and taking into account, i.e. eliminating closed dead-end holes and virtual leaks by avoiding big contacting surfaces or using, for example, pumping holes in case of tight packed parts etc. An exhausting of the SAFEM is done by a rotary pump (RP) for the first low vacuum stage, a turbo molecular pump (TMP) for the second high vacuum stage and a combination of ion getter pump (IGP) and titanium sublimation pump (TSP) for the third UHV stage.

All construction plans of the SAFEM were developed by means of the 3D mechanical CAD software package SolidWorks [Sol]. In order to ensure sufficient mechanical stiffness of the microscope, a mechanical load simulation of some inner parts were performed by means of COSMOSXpress mechanical simulation software which is a part of the SolidWorks package. An example of such simulation of the mechanical load on the main support plate of the SAFEM, holding 3D sliding stages, Macor® insulator and the anode holding system is shown in the Fig. 3.16.

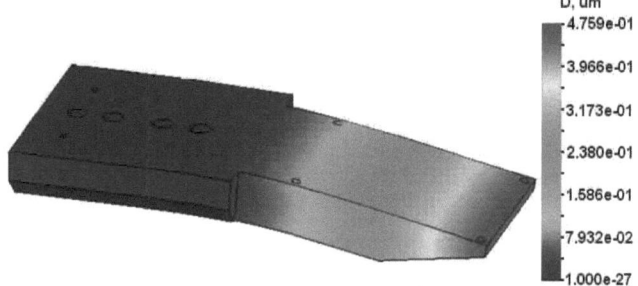

Fig.3.16: Simulation of the mechanical load on the support plate of the SAFEM showing maximum displacement D of 0.476 µm at maximum load of 3 kg resulting in 0.3° tilt of the anode plane with respect to the cathode. The tilt is to be software-wise corrected during the initial cathode tilting procedure.

In order to achieve the required vacuum in reasonable time, the whole system finally has to be exposed to around 200°C. Because of different coefficients of thermal expansion of various materials, constructions of adjacent parts were done in a way eliminating thermally driven mechanical deformations by keeping them in the elastic range and avoiding plastic deformations. A special attention in this instance was made in coupling sites between the Macor® insulator and the metallic parts. Depth of the threads in the insulator and geometry of the insulator and metal parts were adapted to avoid a break of the insulator by heating and cooling.

Final properties of the SAFEM will be:

• direct transfer of photocathodes between preparation chamber, the SAFEM and the transport box, which deliver fresh cathodes to the gun,

• 10^{-10} mbar working pressure,

• optical electrode gap control down to 5 µm resolution,

- imaging of potential emitter distribution over the photocathode surface, contact area and a part of the back plane,
- 25 x 25 mm² scanning range,
- 1 μm scanning resolution
- detailed FE investigations of individual emitters
- up to 10 kV (± 0.1 V) applied voltage (with ms fast PID regulation), corresponds to 200 MV/m macroscopic electric field at around 50 μm gap,
- ± 1 fA current reading resolution,
- fully automated measurement control based on PC/LabView.

First vacuum test of the SAFEM suffered from a strong gassing of the Micos sliding stages. After replacement of some parts and complete cleaning of them by Micos Company, there was a significant improvement of the vacuum. The specified value of 10^{-10} mbar, however, could not be reached still (see Fig. 3.17). Glued end-switches of the motors and the unsuitable lubricant used for ball bearings found to be the major sources of the gassing. As a result, the mass spectra showed presence of a big amount of hydrocarbons in the chamber (Fig. 3.18). Afterwards the sliding stages had been sent back to the Micos, where the end-switches have to be replaced and the lubricant has to be either replaced or completely removed. Actually, the stages are back at DESY and are under the vacuum tests. The results will be known is the nearest future.

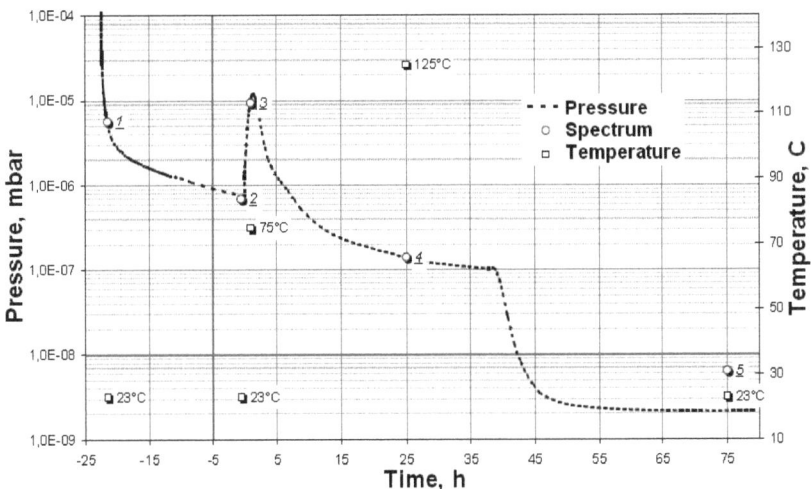

Fig. 3.17: Exhaustion of a chamber with two MTS-65 sliding stages after the first repair.

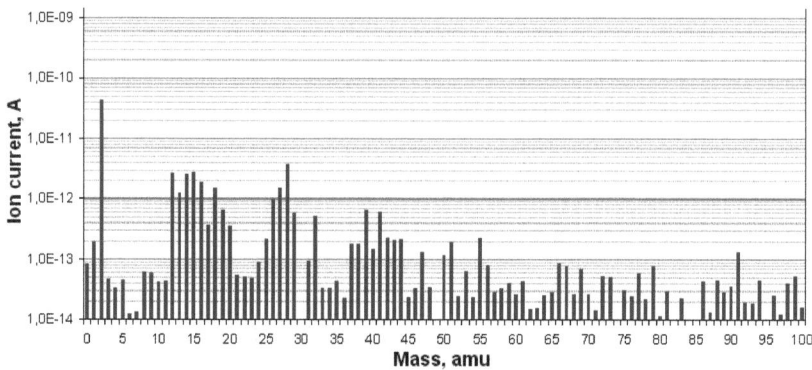

Fig. 3.18: Mass spectrum in a chamber with two MTS-65 sliding stages after the first repair taken at 24 °C and 2.4e-9 mbar (the spectrum is taken in point 5 in Fig. 3.17).

After the successful vacuum tests and first commissioning of the SAFEM, it will be finally included into the preparation and quality control set-up.

In the already planned upgraded stage, the SAFEM will be equipped with the anode exchange system. This will give the opportunity to change resolution of the SAFEM by choosing between 9 anodes of different size without breaking the UHV conditions.

3.4. Surface analysis

3.4.1. Scanning electron microscope (SEM) and EDX

The morphology of typical emitting nanostructures and defects before and after the FE measurements was investigated by a SEM (for example, Phillips XL30 at GSI and Phillips XL30S at the University of Wuppertal). The SEM provides topographic details of a surface and operates by magnetically scanning a focused electron beam, typically from 1 to 40 keV in energy, across the surface of a sample in the high vacuum. The incident electrons, upon interacting inelastically with the sample material produce secondary electrons. By measuring the response of a secondary electrons detector as the beam is scanned over the sample, an image is constructed. Electron beam sources are either thermal emission from heated W or LaB_6 filaments, or FE. The sample in the SEM needs to be at least slightly conducting. Insulating samples are charged up, resulting in a beam deflection and distortions of images. The SEM resolution is limited by a beam spot size (can be as small as 1 nm), a beam jitter, and a spread of a secondary electron yield. Two common additions on the SEM are detectors for backscattered electrons and x-rays. As both the quantity of back-scattered electrons and the spectra of emitted x-rays significantly depend on the charge number of the electrons in the sample, these techniques are used for the elemental analysis. Energy-dispersive x-ray spectroscopy (EDS or EDX) is an analytical technique used for the elemental analysis or chemical characterization of a sample. It is one of the variations of x-ray fluorescence spectroscopy, which relies on the investigation of a sample through interactions between electromagnetic radiation and matter, analyzing x-rays emitted by the matter in response to being hit by charged particles. Its characterization capabilities are due in large part to the fundamental principle. Each element has a unique atomic structure allowing x-rays, which are characteristic of an element's atomic structure, to be identified uniquely from one another. The EDX spectroscopy was used for analysis of defects on Nb surfaces and was performed on the Phillips XL30S SEM at the University of Wuppertal.

3.4.2. Optical profilometer (OP) and AFM

For the measurement of the surface roughness and the localization and characterization of defects, a commercial optical profilometer MicroProf® [Frt] (see Fig. 3.19) was used. It is mounted on a solid granite support plate with an active vibration damping system and placed in front of a laminar airflow system (class 5 ISO) for cleanroom-like conditions. The measurement system combines a small CCD camera for fast orientation with the OP based on a spectral reflection (chromatic aberration) of white light and an AFM in calibrated positions (1 µm precision). Surface profiles of flat as well as curved samples up to 20×20 cm^2 size and 5 cm height difference are measured fast (100×100 pixels in about 1 minute) and non-destructively down to 2 µm lateral and 3 nm vertical resolution. Further zooms into localized defect areas down to 3 nm lateral resolution are performed with the AFM with 34x34 µm^2 scanning range.

Fig. 3.19: MicroProf® OP with AFM.

4. Fabrication and FE results of structured cathodes
4.1. Arrays of CNT columns and blocks

At present CNT and nanofibers are the most prospective cathode materials for triode applications [Xu05] because of their strong FE at low electric fields [Hee95, Rin95]. Applications such as flat-panel displays [Kum10, Yi02], multiple electron beam lithography [Kim06b, Koj10], x-ray sources [Lee10, Sch10b] etc. require high emitter number density with uniform and stable FE properties at fields well below 50 V/μm. Furthermore, effective current densities above A/cm^2 are the challenge for microwave power generation and amplification devices [Bow02, Bro10, Har09, and Lin09]. The FE homogeneity of flat CNT cathodes, however, is still limited by their rather fast and uncontrolled growth [Din06] which usually leads to strongly varying field enhancement and current carrying capability of the individual emitters [Dea01, Sem02] in defined FE arrays [Hsu05, Man05, Lee05]. The patch arrays of the aligned carbon nanofibers can be fabricated onto catalyst patterns by the chemical-vapor deposition (CVD) of them from hydrocarbons but suffer from limited efficiency [Fan99, Teo01]. Sparsely grown non-aligned but well-anchored CNT synthesized by the ferrocene CVD in the porous alumina templates resulted in high emitter number and current densities [Lys07], but rather brittle cathodes of irregular shape made them unsuitable for triode applications. Therefore, patches with multiple emitters and reduced mutual shielding might provide a suitable strategy to improve the homogeneity and current stability of the CNT cathodes. In this work, systematic results on the FE investigations of structured arrays with different columns and blocks of entangled CNT are reported. The columns were preferentially grown on flat Si substrates applying atmospheric pressure CVD technique with a volatile catalyst [Lab09a]. The blocks were grown by a water assisted CVD with patterned catalyst [Jos10a, Jos10c]. Spatially resolved as well as integral FE measurements of the test array cathodes with the CNT structures demonstrate both high efficiency and high current capability especially for long embedded CNT columns [Lab09b, Nav10a, Pru09a, and Pru09b]. Long-term current stability and gas exposure tests have been done and will also be presented and discussed.

The FE properties of the structured CNT cathodes were investigated with the FESM and IMLS. After the tilt correction of the cathodes, the Ø30 μm anode was used for the non-destructive imaging of the potential emitter distribution [Dan06] over the whole or arbitrary chosen area of the structured cathodes. In addition, small scan measurements on selected patches were made with the needle anodes (tip radius 8 µm or 3 µm). Finally, local FE measurements of individual blocks or columns were carried out with suitable anodes.

The integral properties of the structured CNT cathodes were determined in dc and pulsed modes at the base pressure of 2×10^{-7} mbar, as well as the long-term performance of the CNT cathodes was tested in dc mode at the base and under enhanced nitrogen or oxygen pressure levels up to 3×10^{-5} mbar.

4.1.1. Preferential CVD synthesis of the CNT columns on Si

The method was developed at Belarusian State University of Informatics and Radioelectronics (BSUIR) [Bsu] by the group of Prof. V. A. Labunov including A. Prudnikava, J. Shaman and B. Shulitski. The columns were preferentially grown on flat patterned n-Si/SiO$_2$ substrates applying the atmospheric pressure CVD technique with a volatile catalyst as follows: low-resistivity flat silicon (n-Si, 4.5 Ωcm) substrates of 6x6 mm^2 size have been oxidized at 1100°C in oxygen atmosphere. The resulting SiO$_2$ oxide layer of 0.3 µm thickness have been structured by means of the optical photolithography and the wet selective chemical etching as shown in the Fig. 4.1.

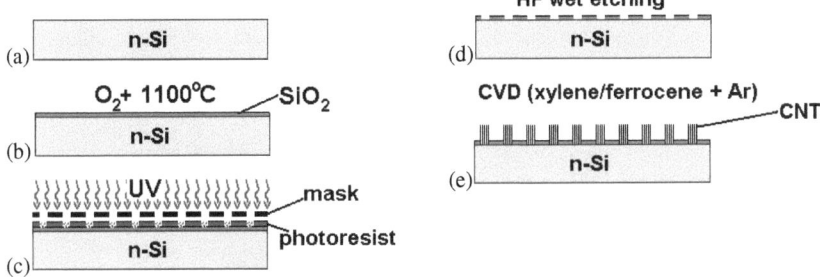

Fig. 4.1: Fabrication steps of structured CNT cathode on Si substrate.

Finally, CNT have been grown on the structured substrates by the atmospheric pressure CVD method with a ferrocene/xylene mixture as volatile catalyst source. Preferential CNT growth on the bare Si surface in the opened windows of the SiO$_2$ layer was achieved by adjusting parameters of the synthesis process, i.e. the concentration of the catalyst in the feeding solution, the gas-carrier flow rate, the temperature in the reaction zone, etc. as described elsewhere [Lab09a]. Growth times of 30 s and 2 min were chosen to obtain multiwall CNT of different length. Figure 4.2 shows resulting structured arrays of the CNT columns which form 4 quadrants of 2x2 mm^2 size with round patches and pitch to diameter ratio of 160/50, 100/30, 100/50 and 100/10 µm, respectively. The morphology of the fairly uniform CNT columns was further investigated by SEM (JEOL JEM-100-CX). Vertically aligned columns of about 20 µm and 50 µm net heights consisting of densely packed entangled CNT are revealed and shown in the Fig. 4.3. The columns are embedded in a cloud-like floor of shorter CNT still grown on the SiO_2. At the top of the columns some multiwall CNT of less than 100 nm diameter and up to 5 % over length are exposed. Considering the factor four of the chosen growth times, the background CNT layer might have a thickness of a few µm for the short and up to 30 µm for the long columns. Therefore, an improved electrical contact and mechanical stability can be expected for the long columns.

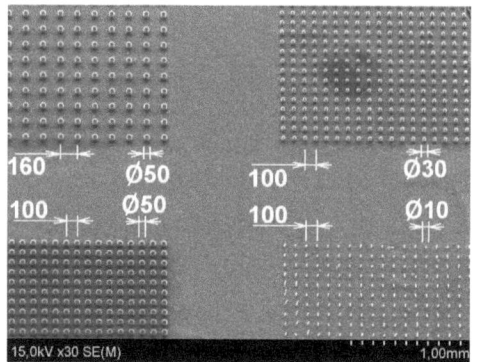

Fig. 4.2: SEM image of the test arrays of the CNT columns on the Si substrate (sizes given in µm).

Fig. 4.3: Some SEM side-view images of short (20 µm) and long (50 µm) columns of 30 µm diameter consisting of entangled multiwall CNT (Ø < 100 nm) embedded in the floor of shorter CNT.

4.1.2. Water assisted CVD synthesis of the CNT block arrays

The water assisted CVD method of the CNT synthesis was recently introduced at Technical University of Darmstadt (TUD) [Tud] by the group of Prof. J. Schneider including J. Engstler and R. K. Joshi. Structured arrays of aligned CNT blocks were fabricated on flat Si substrates by applying a bimetallic (aluminum and patterned iron) catalyst and water assisted CVD [Jos10c]. Light-doped p-type silicon <100> was used as the substrate. The substrates

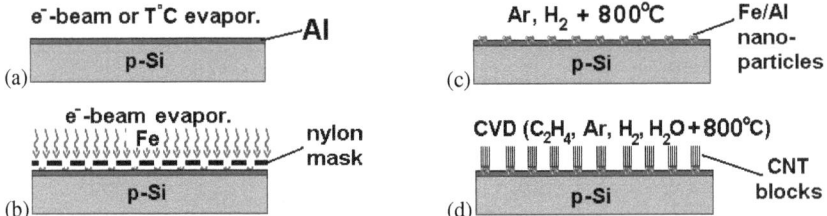

Fig. 4.4: Fabrication steps of the structured CNT cathodes on the Si substrate.

were covered by 10 to 12 nm thick aluminum buffer layer (Fig. 4.4), deposited using the thermal evaporation (heating of Al in the boron nitride crucible using a tungsten filament) or electron beam evaporation. Afterwards, patterned iron film of 0.6 to 1.5nm thick was sputter deposited using electron beam evaporation. Patterning of the iron was achieved by using the nylon meshes of different shape resulting in various block array structures. The mesh was placed on the aluminum layer prior to the iron deposition and removed afterwards. Finally, the samples were placed inside of a Ø2.5 cm quartz-tube-oven and heated up to 750°C under flow of argon and hydrogen (Fig. 4.5). At this temperature, Al and Fe films crack, recombined and formed inter-metallic phase and small Fe/Al nanoparticles. The nanoparticles served as catalyst for the CNT growth happening by a root growth mechanism. For the CNT growth,

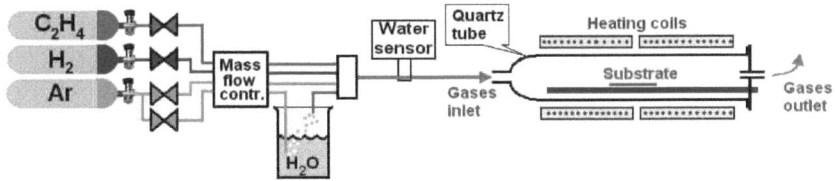

Fig. 4.5: Schematic sketch of the water assisted CVD setup at TUD.

flow of ethylene and small amount of water vapor started through the oven. The water vapor was picked up by a bubbling of small amount of argon through a water bubbler and acted as a catalyst activator. Typical growth rate of the CNT was about 20 to 25 µm per minute. For growth times of 5 to 10 min, the resulting multiwall CNT formed vertically aligned arrays of uniform rectangular blocks of 50, 100 or 140 µm width (w), 150 to 600 µm height (h), and 230, 250 and 300 µm pitch (p) (see Table 4.1) as partially shown by SEM images in the Fig. 4.6. Nearly 100% of selectivity of the CNT growth was achieved showing the possibility of this technique to fabricate regular CNT block arrays of various geometries.

Table 4.1. Overview of the structured CNT cathodes investigated here.

Cathode		#1	#2*	#3	#3*	#4	#4*
Type	pure	x		x		x	
	TiO$_2$ coated		x		x		x
Width [µm]		140	140	100		50	
Height [µm]		600	600	150		200	
Pitch [µm]		230	230	250		300	

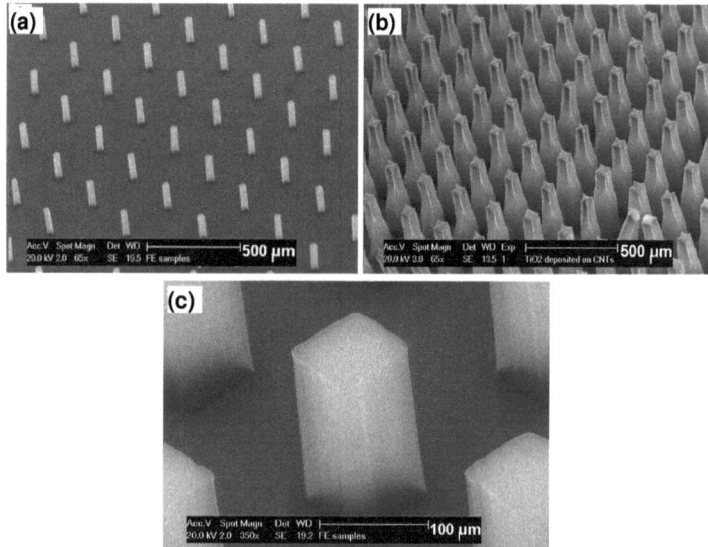

Fig. 4.6: Typical SEM images (45° view) of the structured arrays: a) pure CNT (w = 50 µm, p = 300 µm, h = 200 µm); b) TiO$_2$-coated CNT (w = 140 µm, p = 230 µm, h = 600 µm) and c) pure CNT (w = 100 µm, p = 250 µm, h = 200 µm) blocks.

Some of the CNT block arrays were CVD-coated with a thin TiO$_2$ layer (< 50 nm) to study current stabilization effects and influence of it on the onset field and the maximum current capability of the blocks. A liquid precursor of the TiO$_2$ was evaporated and carried into the CVD reaction zone of the oven by means of an argon gas flow as described elsewhere [Jos10a, Jos10b]. It is remarkable that the TiO$_2$-coated blocks are sharpened and partially tilted due to their high aspect ratio.

4.1.3. Efficient high-current FE from arrays of CNT columns

Low-resolution FESM maps of the whole structured CNT cathode with short columns (Fig. 4.7) confirm a fairly homogeneous emission from the patch array regions, but less pronounced FE occurred between the arrays as well.

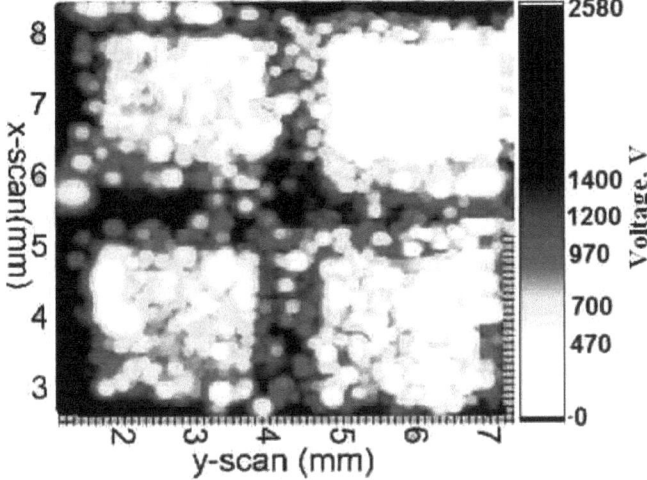

Fig. 4.7: Regulated voltage map for 1 nA FE current of structured cathode with 4 arrays (2x2 mm^2) of short (20 µm) CNT columns corresponding to the Fig. 4.2 (scanned area 36 mm^2, low resolution anode Ø 30 µm, gap 40 µm).

Typical medium resolution FESM maps of about 0.5 mm^2 scan area within the four different quadrants of this cathode shown in the Fig. 4.8 demonstrate that well aligned and efficient FE from nearly 100% of the patches was achieved at electric fields below 15 V/µm. These FESM maps, however, reveal the presence of some emitters between the patches either. Moreover, multiple emitters especially for 50 µm diameter columns are clearly visible in the medium resolution (e.g. Fig. 4.8(c)) and other high-resolution FESM maps. Typical low and high-resolution FESM maps of the structured CNT cathode with long columns are shown in the figure 4.9. Accordingly, most patches emit already at electric fields below 10 V/µm, but some pronounced FE occurred again in the area between four arrays and between the columns resulting in less aligned emission (Figs. 4.9(b)-4.9(c)). In comparison, structured cathodes with longer CNT columns are more inhomogeneous and less aligned than those ones with shorter columns. It is due to thicker CNT floor that eventually causes strong FE, too. Therefore, the selectivity of the CNT growth process should be further improved especially for triode devices requiring low gate currents.

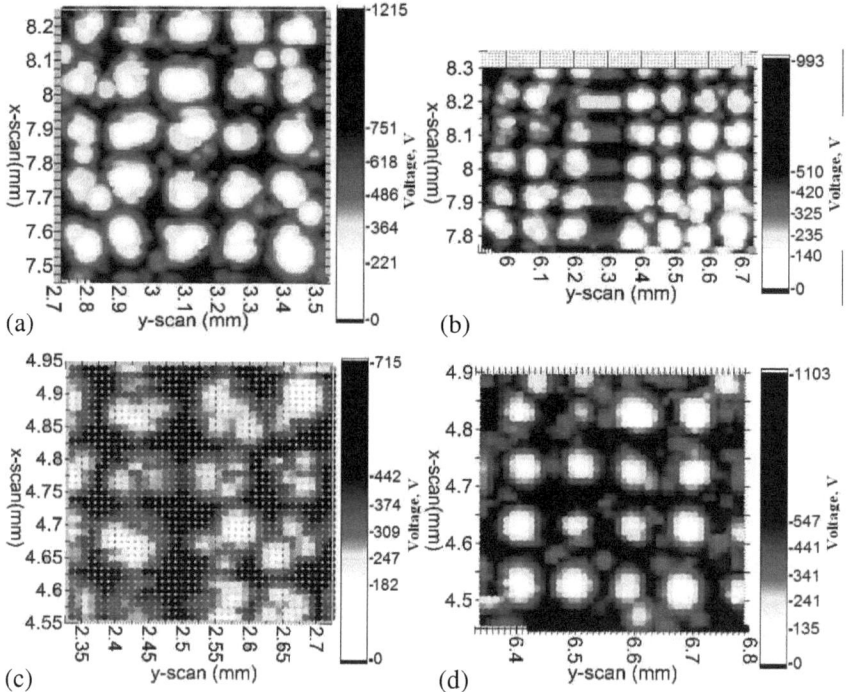

Fig. 4.8: Medium resolution regulated voltage maps (for 1 nA, anode Ø 8 µm, gap 20 µm) of the same structured cathode as in Fig. 4.7 (20 µm CNT columns): a) upper left (0.64 mm²), b) upper right (0.44 mm²), c) lower left (0.2 mm²) and d) lower right (0.64 mm²) quadrant.

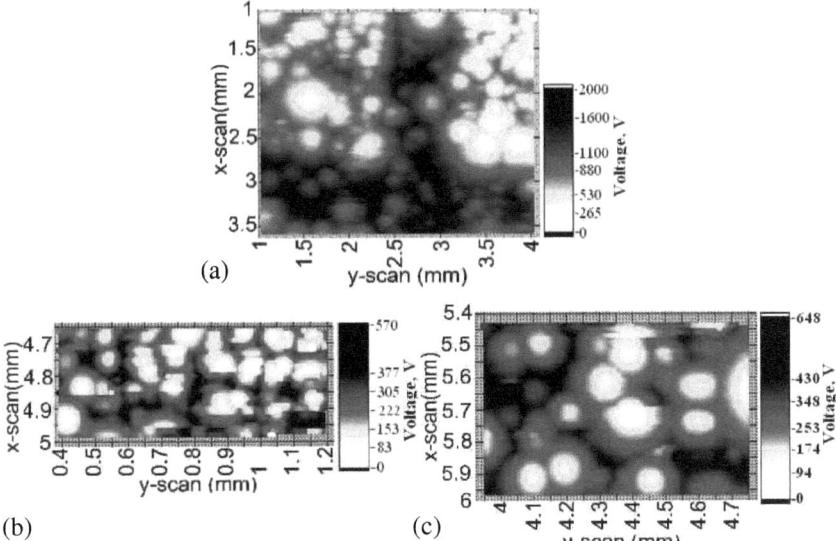

Fig. 4.9: Regulated voltage maps (for 1 nA) of structured cathode with long (50 μm) columns: a) two upper quadrants corresponding to Fig. 4.2 (scanned area 7.8 mm², low resolution anode Ø 30 μm, gap 60 m); b) lower left (0.28 mm²) and c) lower right quadrant (0.64 mm²) in high resolution (anode Ø 3 μm, gap 15 μm).

Integrally measured I-V curves of single patches with the Ø 30 μm anode showed rather stable FN behavior up to currents of 80 μA for short and 500 μA for long CNT columns at electric fields up to 30 V/μm as shown in the Fig. 4.10 (a). It is remarkable that the current carrying capability of the short columns, i.e. 20, 40 and 80 μA, strongly increased with their diameter (10, 30 and 50 μm, corr.), while the long ones were only limited at 0.5 mA by the maximum range of the ammeter. This surprising result with respect to current limits of single MWNT of about 20 μA [Dea01, Sem02, and Nil00] might be explained only by contribution of many entangled CNT emitters to the patch current and by their improved embedding in a highly conductive CNT floor, especially for the long columns. In Fig. 4.10(b) the electric field enhancement factor β_{eff} of randomly chosen columns of both lengths are compared (assuming a work function of 4.9 eV for the CNT [Grö01]). Short CNT columns provided a larger spread of the β_{eff} and the electric onset field values, while the long ones better confirmed the hyperbolic correlation expected from the FN theory for a given current. Mean values of β_{eff} of 168, 111 and 185 (400) resulted for short (long) CNT columns of 10, 30 and 50 μm diameter, respectively. These β_{eff} values cannot be explained by the ratio of the column height and the CNT radius but hint for varying mutual shielding effects within the columns and field enhancement by outstanding individual CNT.

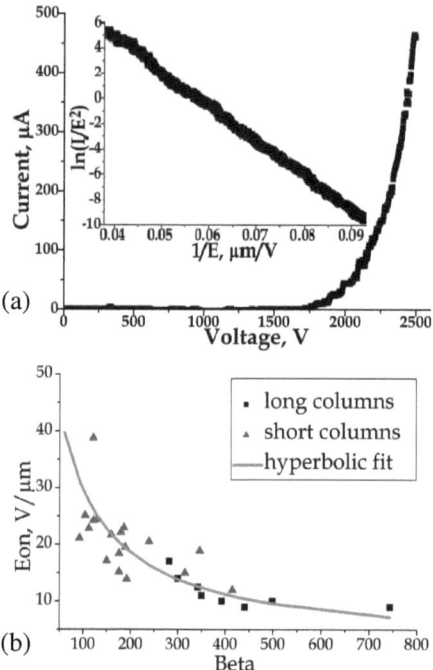

Fig. 4.10: (a) Typical I-V and corresponding FN plots (inset) of a single CNT column (Ø 50 μm, 50 μm net height) measured with the tip anode of Ø 30 μm showing excellent stability up to the maximum current of 500 μA. (b) Onset electric field E_{on} (for 1 nA) vs. field enhancement factor β_{eff} plot for both length of Ø 50 μm CNT columns showing expected hyperbolic correlation.

The integral performance of the same structured cathodes with short and long CNT columns was tested with the IMLS. In the Fig. 4.11, the current distribution over the cathodes obtained in dc and pulse mode is displayed by the luminescent screen images. Relatively good FE current homogeneity is visible already at electric fields below 10 V/μm in all four quadrants, i.e. independent of the pitch to patch ratio of the arrays. It should be mentioned that this result was achieved only after some pulse mode processing of initially strong emitters at the edges of the cathodes. Obviously, the light spots corresponding to resolvable emitters (20 μm) at low field levels (Fig. 4.11(c)) smear out for higher currents due to the bright halo produced within the luminescent screen (Fig. 4.11(d)).

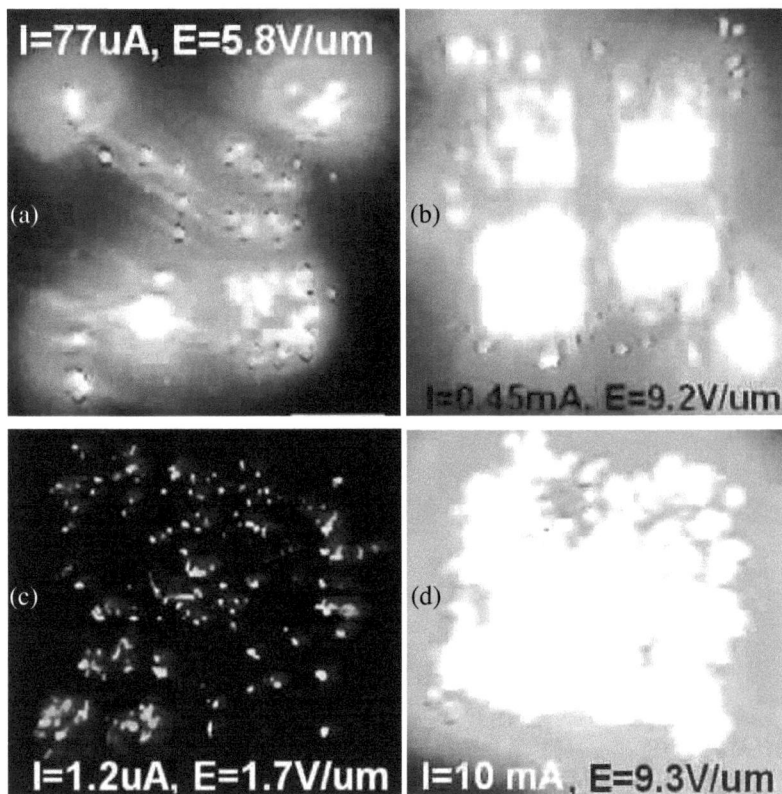

Fig. 4.11: The luminescent screen images of the structured cathode with short (a, b) and long (c, d) CNT columns (arrays position correspond to Fig. 4.2) in dc (a, c) and pulse modes with duty cycles of 2:20 ms (b) and 1.5:150 ms (d).

The maximum achievable cathode current of 81 (98 µA) in dc and 2.5 (10) mA in pulse mode and for short (long) columns was always limited by the power load to the luminescent screen which caused some sulfur evaporation (as detected by the mass spectrometer) and risks successive cathode destruction by plasma discharges. Nevertheless, achieved pulsed current density of at least 1.8 (40) mA/cm^2 for the cathodes with short (long) CNT columns can be derived. For the full exploitation of optimized CNT cathodes, therefore, metallic anodes will be required.

Long-term stability tests have been carried out for the cathodes with both short and long CNT columns in dc mode for some hours under various vacuum conditions. As expected, much better stability was obtained for the longer CNT columns as shown in the Fig. 4.12(a).

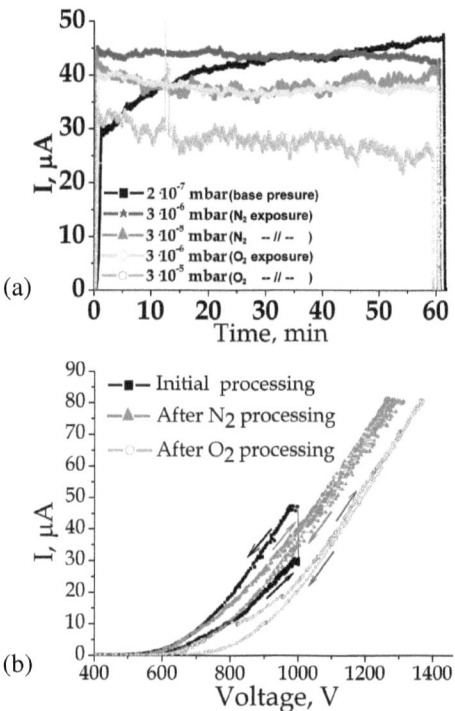

Fig. 4.12: (a) Integral long-term current stability of the structured cathode with long CNT columns at constant supply voltage (1 kV, 2.1 MΩ ballast resistor) measured at base pressure (2×10^{-7} mbar) and at enhanced partial pressures of N_2 and O_2. (b) The integral I-V curves (300 μm gap) at base pressure showing processing effects for different gases (arrows indicate up- and down voltage cycles).

At the base pressure of 2×10^{-7} mbar a strong activation of emitters was observed especially during the first 20 minutes which can be explained by the desorption of gases and/or electric field alignment of CNT [Bon03]. In contrast, the integral FE current decreased slightly under nitrogen and strongly under oxygen exposure, which was done in four steps (at 10^{-6} mbar, 3×10^{-6} mbar, 10^{-5} mbar and 3×10^{-5} mbar) for one hour each. The effects summarized in Fig. 4.12(a) can be probably explained by passivation due to adsorption of gas molecules (reversible effect), bombardment of the CNT with ions leading to their partial destruction or burning of emitting CNT tips, especially in case of oxygen (irreversible effects). Various ions might also activate the columns by splitting of the CNT bundles, by sharpening of their tips or by removing adsorbates [Xih09, Kwo07, and Bor07]. The current spike in the I-V curve for the highest O_2 pressure (Fig. 12(a)) was caused by a local spark as suggested by the corresponding IMLS images. Since the irreversible effects are mainly caused by oxygen, operation during 1 hour at 3×10^{-5} mbar with O_2 gas would correspond to about 1000 hours at the base pressure of 2×10^{-7} mbar with 35.3% of H_2O as a main source of oxygen. In Fig. 4.12(b), the permanent processing effects of both gases are compared by means of the integral I-V curves of the cathode. While the hysteresis effects at low fields confirm the activation and passivation of emitters by the desorption and adsorption of gases, the partial destruction of them leads to shift to higher fields for the same current and occurred only for the oxygen processing.

The balance between passivation and activation of emitters was always monitored by the IMLS imaging. It has revealed short-term current fluctuations for less than 20 % of the spots and their partial redistribution at comparable current levels. In order to demonstrate the whole current processing and gas exposure effects on the cathodes, averaged consecutive IMLS images (10 over 1 s) of the initial and a final status have been subtracted by means of the image analysis software AnalySiS®. The result for the cathode with long CNT columns for the integral current of 10 µA is shown in the Fig. 4.13. After the full processing time, 12% of the emitting area remained stable, while 44.3% (43.6%) has been activated (deactivated). Considering the moderate field levels chosen to avoid the halo effects mentioned above, this result supports our strategy to improve the homogeneity of structured cathodes by multiple CNT emitters per patch.

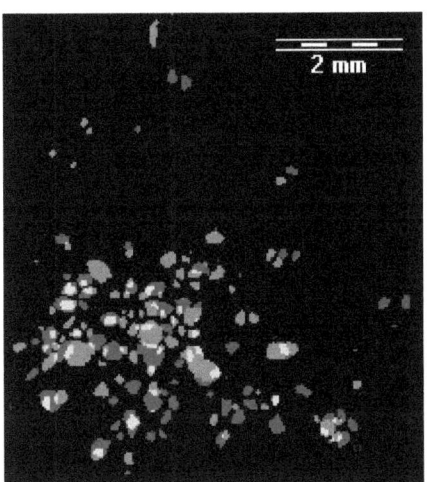

Fig. 4.13: Software analysis of the IMLS images obtained for 10 µA integral current initially (at 785 V) and finally (at 900 V) for the same cathode as in Fig. 4.12 showing the processing effects of both gases. The emitter distribution has partially changed resulting in stable (yellow), activated (green), and deactivated (red) emitters.

In conclusion, structured arrays of entangled CNT columns embedded in a floor of short CNT were successfully fabricated by preferential CNT growth on bare Si as compared to SiO_2 layer. Cathodes with short CNT columns showed well-aligned and efficient FE from nearly 100% of the patches at electric fields below 15 V/μm but only moderate patch currents up to 80 μA and might be useful for triode applications. In contrast, patches with long CNT columns yielded stable currents of at least 0.5 mA suitable for power devices. These very high current values as well as the derived field enhancement factors confirm the presence of multiple CNT emitters per patch. Rather homogeneous cathodes with long columns provided the integral current density of up to 40 mA/cm^2 at 10 V/μm in pulse mode which was only limited by the power load to the luminescent screen. The long-term stability of the cathodes was proven by dc processing under enhanced oxygen pressure. Combining reproducibly high and stable patch currents with the obtained FE homogeneity, current densities in the range of A/cm^2 can be expected for optimized cathodes with arrays of long CNT columns of 50 μm diameter, 160 μm pitch and mm^2 size.

4.1.4. FE properties of aligned pure and TiO$_2$-coated CNT block arrays

The FE properties of various cathodes with aligned pure and TiO$_2$-coated CNT blocks arrays were systematically investigated by the FESM using tungsten tip anodes of adjusted size, i.e. with a diameter $Ø_a$ and a gap Δz in the order of block width w. Fig. 4.14 demonstrates that high FE efficiency from all CNT block arrays was obtained at electric onset fields E_{on}(1 nA) ≈ 10 V/μm. Good alignment of emitters was achieved depending on the pitch to block width ratio and is the best for the wide-spaced CNT block arrays. The TiO$_2$-coated CNT block arrays have shown very similar results in terms of alignment and efficiency as the pure ones. Moreover, multiple emitters per block are clearly visible at least for the large (w > 100 μm) blocks and surely contribute to the high efficiency of such cathodes. Different E_{on} and β_{eff} values of the cathodes correspond to the geometry of CNT outgrowth rather than to the height of the blocks [Nav10c].

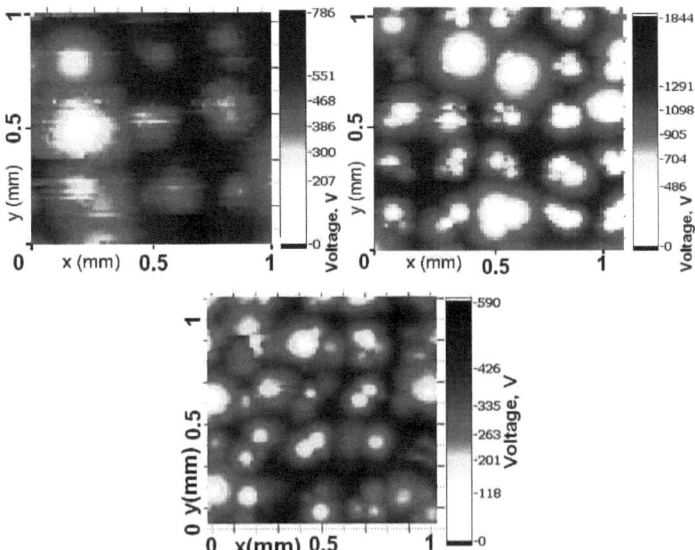

Fig. 4.14: Regulated voltage maps (area ~ 1 mm^2, $\emptyset_a$ = 30 µm, 1 nA fixed FE current) of the structured arrays of the same pure CNT (a, $\Delta z \approx 50$ µm), TiO_2-coated CNT (b, $\Delta z \approx 30$ µm) and pure CNT blocks (c, $\Delta z \approx 50$ µm) as shown in Fig. 4.6.

In order to investigate current stability as well as maximum current capability of the cathodes, integral measurements of single blocks were performed with $\emptyset_a$ = 30 or 100 µm anodes after field calibration by means of the V(z)-plots. After some initial processing effects and low current fluctuations, stable currents up to 300 µA at 37 V/µm have been achieved for some pure and up to 100 µA at 32 V/µm for TiO_2-coated CNT blocks as shown in the Fig. 4.15. The high current capability of individual blocks obtained here confirmed multiple CNT emitters per block. Current densities in the range of A/cm^2 as required for high power devices could be expected for the CNT block arrays in case of sufficient FE homogeneity.

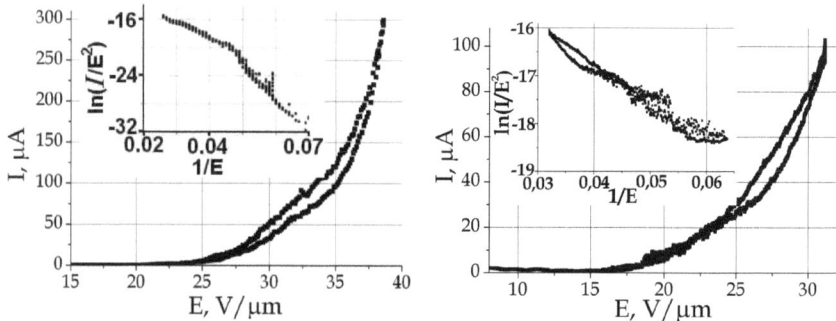

Fig. 4.15: Current-field curves and FN plots (insets) of single pure (w = 140 µm, p = 230 µm, h = 300 µm, left) and TiO$_2$-coated CNT blocks (w = 100 µm, p = 250 µm, h = 200 µm, right) showing maximum stable current limits.

Often-observed midfield hysteresis can be caused by an effective gap change due to reversible alignment of single CNT or even clusters from the blocks. Most CNT blocks, however, showed irreversible current instabilities in the range of 10 - 200 µA due to stepwise CNT disruption and complete block destruction at 200 - 800 µA.

In order to investigate TiO$_2$ coating effects, the cathodes N3 and N4 were investigated before (with pure CNT) and after TiO$_2$ coating (N3*, N4*) to ensure the same geometry of blocks, quality of the CNT, contact interface etc. A slight change of the onset field was observed as well as the field enhancement by TiO$_2$ coating (Fig. 4.16) but the big scatter of the data was still typical for both types of the cathodes. The increase of the field enhancement factors (from 496 to 639 mean values) and corresponding decrease of the onset field (from 16 to 7 V/µm mean values) can be result of the observed sharpening of the blocks during CVD deposition of TiO$_2$. The scatter of the data is resulting from the limited geometrical uniformity of the blocks, especially because of their height and rectangular cross-section. Such cross-section of the blocks leads to inhomogeneous distribution of the electric field on the upper edges. The field is more enhanced at the corners as can be partially seen on the corresponding voltage maps in the Fig. 4.14. Therefore, height of the blocks should be reduced to about 20 µm to avoid their tilting and difference in cross-section and height. The pitch of the arrays should be at least four times of the block width and three times of their height in order to improve alignment and decrease their interference and mutual shielding.

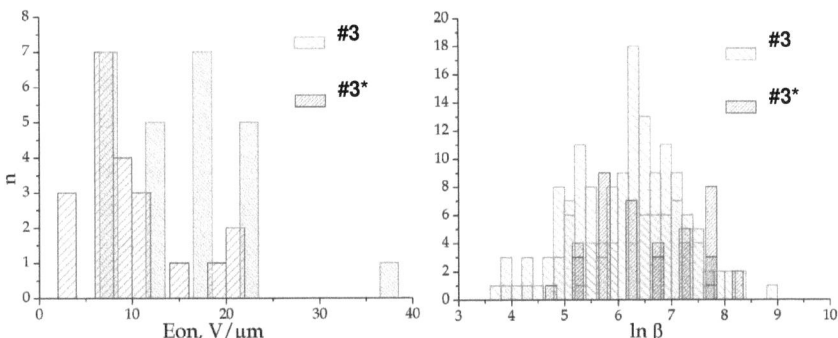

Fig. 4.16: Histograms of E_{on} and β factors of all measured blocks on cathodes #3 with pure and #3* with TiO^2 coated CNT blocks.

There is some evidence that TiO_2 coating might stabilize FE at low currents (I < 10μA) (Fig. 4.17) showing less stepwise deactivation of emission and more FN like behavior

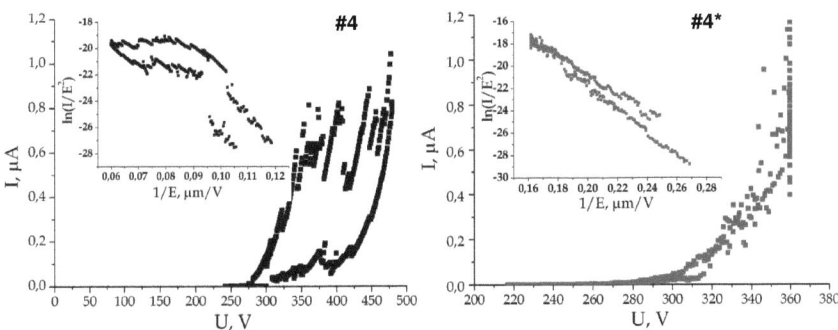

Fig. 4.17: Typical current-field curves and FN plots (insets) of single pure (left, #4) and TiO_2-coated (right, #4*) CNT blocks of the same cathode at currents below 10 μA

with straight FN plots. It can be explained by smoothing of the blocks and binding of individual (probably even loosen) CNT on the blocks by the TiO_2 layer. However, it was discovered also that TiO_2 coating lead to factor of 2-3 reduction of maximum currents limits of individual blocks comparing to the pure ones. It can be assumed that at certain current or field values TiO_2 nanoparticles can be evaporated and ionized leading to a plasma discharge and destruction of the blocks at the much lower currents comparing to pure CNT blocks.

The figure 4.17 (left) shows also the often-observed saturation of the FN plot. The saturation is often misinterpreted as an effect of a limited current supply due to intrinsic material properties, an effect of the overestimated real voltage between emitters and anodes by neglecting a voltage drop on the emitters itself, etc. In reality, it is believed that the saturation is due to the stepwise destruction of emitters, as it is seen in the figure 4.17 (left). It leads to such bended curves, which consists of a big amount of small straight FN-like plots. Determination of the field enhancement factor and emission area from such set of FN-like plots as a unified one will lead to unrealistic β and S values (too high and too small corr.). In some cases, however, the above-mentioned effects may also take place and understanding of the saturation require more careful theoretical (including model calculations) and experimental analyses.

It is remarkable that glowing spots at the top of CNT blocks have been often observed especially for TiO_2-coated ones. This obvious local heating was frequently followed by their destruction in case of further current increase and will be further discussed elsewhere [Nav10b].

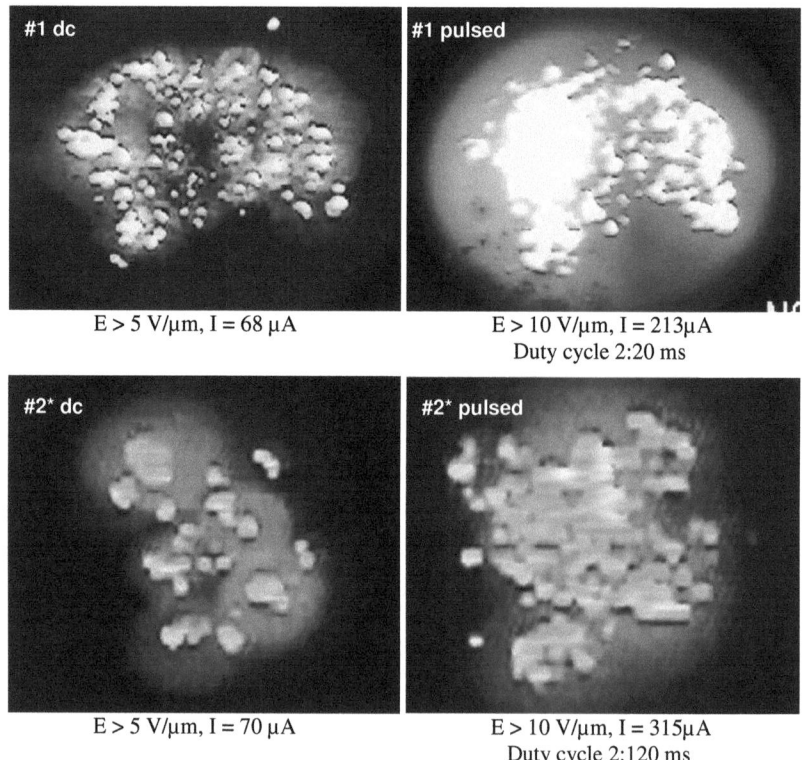

Fig. 4.18: Luminescent screen images of pure #1 (two upper) and TiO_2-coated #2*(two lower) structured cathode with CNT block arrays in dc (left) and pulse (right) modes.

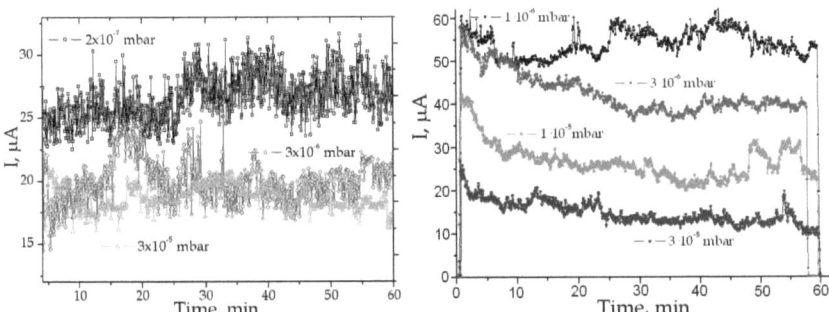

Fig. 4.19: Long-term FE tests of pure (#1, left) and TiO_2 coated (#2*, right) CNT blocks arrays at enhanced oxygen pressure.

The integral performance of the structured cathodes with pure and TiO_2-coated CNT block arrays was tested with the IMLS. In Fig. 4.18, the current distribution over the cathodes obtained in dc and pulse mode is displayed by the luminescent screen images. Most of the CNT blocks are emitting already at electric fields below 10 V/µm, but the cathode homogeneity has to be improved yet.

Long-term stability tests at stepwise enhanced O_2 pressure have been performed for both pure (#1) and TiO_2-coated (#2*) CNT block arrays in the dc mode for about 1 hour each step. As expected, TiO_2-coated CNT cathode has shown slightly better stability with less short-term fluctuations as shown in Fig. 4.19 but in both cases, the integral FE current decreased under oxygen exposure. The test showed, however, factor of two higher sensitivity of TiO_2-coated CNT to oxygen and mainly irreversible FE degradation for both types of the cathodes. The effect can be explained by bombardment of the CNT with ions leading to their partial destruction or burning of emitting CNT tips by oxygen. Studies of the effects of TiO_2 coating on FE performance of CNT cathodes, including oxygen exposure tests, will be done in more detailed way in the nearest future. The current spikes in the I-t curves, observed more often for pure CNT and in a high current (field) range, and were caused by local sparks as suggested by the corresponding IMLS images revealing mechanical instability of CNT and their pulling out in the field. Total processing time of 1 hour of the cathodes at enhanced oxygen pressure (which corresponds to about 1000 hours at the base pressure) indicates ability of the cathodes to work for long time without complete degradation. Another positive aspect of the CNT cathodes is a recovering effect. Because of redundancy and multiple emitters per patch, achievement of a required current value is mostly possible by operation at higher fields (see Fig. 4.20), processing at improved vacuum conditions, short-term activation at higher current/field values etc.

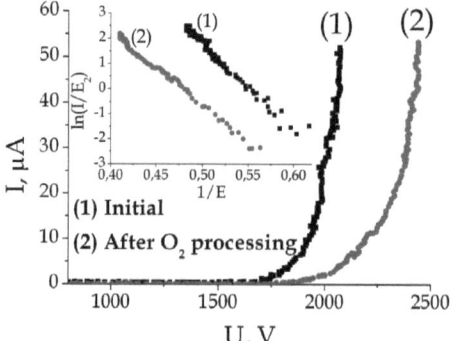

Fig. 4.20: Integral current voltage characteristics of TiO_2 coated CNT cathode showing FN-like behavior (straight line of the inset plots) of the FE and 15% shift to higher field after the O_2 processing.

The water assisted CVD method of the CNT deposition realized at TUD showed 100% selectivity of growth comparing to the atmospheric pressure CVD method from BSUIR. Both types of the cathodes showed high FE efficiency and rather satisfactory alignment from selected regions of the CNT patch arrays obtained at comparable electric fields. Long CNT columns from BSUIR, in contrast, yielded much higher stable currents from an individual column due to better embedding and contact to the substrate. Therefore, improvement of contact interfaces of both CNT types to substrates has to be considered. Current capabilities from individual patches of both types of materials as well as derived field enhancement factors confirm the presence of multiple CNT emitters per patch. The recovering effect due to the redundancy and multiple emitters confirm our concept to improve the FE performance of the cathodes. Integral current density of the cathodes and their long-term performance are very promising. However, the homogeneity over a full cathode has to be improved yet, by adjustment of geometry and better uniformity of the individual patches. Use of the actual cathodes for real application required high quality stable and homogeneous cathodes is not possible, yet. However, some application like simple x-ray sources can be already built with this material.

A more detailed FE study of such CNT cathodes with improved geometry and uniformity of the blocks as well as FE spectroscopy measurements (started in [Bor10]) for better understanding of coating effects are to be done in the nearest future.

4.2. Arrays of gold and platinum nanowire patches

Cold cathodes on the basis of MNW with high aspect ratio (AR) and high conductivity fabricated by the ion-track template technique [Toi01, Vil04] can provide a curious alternative to the CNT with respect to better homogeneity, stability and controllability of the nanostructures in defined FE arrays. The method allows in particular control of the number density of NW by fluence of the ions applied during the template irradiation, whereas the wire diameter and the length are controlled by time of the track etching and electrochemical deposition, respectively [Liu06a]. Based on the promising FE results obtained for randomly distributed copper [Mau06], gold-coated nickel [Dan07b], and gold NW [Dan08] in diode configuration, this study focuses on the development of structured cathodes with defined gold and platinum NW patch arrays which might be suitable for triode applications with low gate currents. It will be reported about FE results on square and triangular arrays of Au-NW and Pt-NW patches with different pitch sizes which were obtained by using a structured metallic shadow mask during the heavy ion irradiation of the polycarbonate template foil. An influence of the number density and aspect ratio of the NW on the morphology of the free-standing solitary or clustered nanostructures of the resulting patch arrays was systematically investigated by means of the SEM and correlated to their FE performance as measured with the FESM [Nav08, Nav09a, Nav09b, Nav09e, Nav10d, Nav10e]. Moreover, the effective field enhancement factor was derived from locally measured I-V curves and correlated to the patch geometry. For selected NW patches the maximum achievable current and resulting changes of the corresponding nanostructures were also investigated. The impact of these results on the possible optimization of NW cathodes for FE triodes will be discussed.

Production of the NW patch arrays was carried out at UNILAC accelerator facility (GSI, Darmstadt), while FE investigations of them were done at University of Wuppertal. High-resolution SEM investigations were done at both institutes.

4.2.1. Ion track template technique

The successive fabrication steps of structured MNW cathodes are shown in the Fig. 4.21. First of all, the templates were produced by exposing 30 µm thick polycarbonate foils (Makrofol N, Bayer AG) through a ~200 µm thick metallic mask (brass MS63 for square and Invar for triangular arrays) to energetic heavy ions. The irradiation experiments were performed at the UNILAC using 2.6 GeV U ions and fluencies of 10^6, 10^7, or 10^8 cm^{-2}.

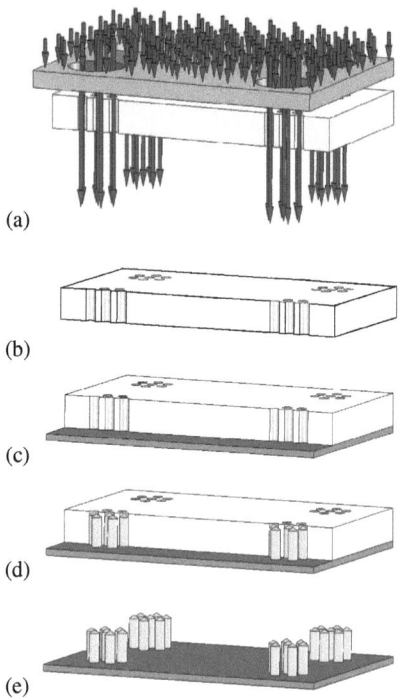

Fig. 4.21: Illustration of the structured cathode fabrication: a) irradiation of polycarbonate foil (white, cut is shown for clarity) with heavy ions (red arrows) through the metallic mask (grey); b) chemical etching of ion tracks; c) deposition of the metallic backing layers (orange); d) electrochemical filling of the pores with Au or Pt nanowires (yellow); e) dissolution of polycarbonate template.

A test mask with four quadratic arrays of patches of 50 µm diameter and different pitch size (100 and 150 µm) (see Fig. 4.22) was used to optimize the fluence (10^6-10^8/cm^2) and NW length (7-28 µm) with respect to emitter efficiency and alignment [Nav09a]. The second mask with a triangular array of 150 µm diameter patches and 320 µm pitch (Fig. 4.22) was then employed to develop cathodes suitable for triode devices.

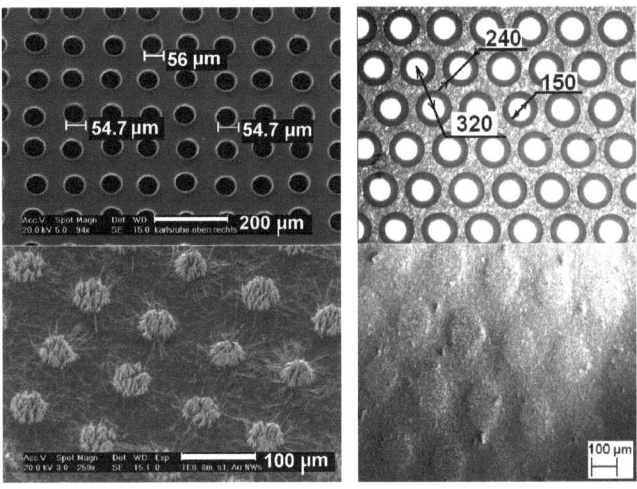

Fig. 4.22: SEM and optical microscope images of the metallic masks with: Ø50 µm cylindrical holes in a square array with 100 µm pitch (left) and conical holes (upper Ø240 µm, lower Ø150 µm) in a triangular array with 320 µm pitch (right) and resulting arrays of NW patches (Ø50µm left and Ø180µm right, 45° and 25° view angles corr.).

The ions create damage zones in the foils, which were selectively etched (in 6M NaOH at around 50°C) as described in more detail elsewhere [Toi01]. Resulting polycarbonate templates provide regular patch arrays (10000 or 4444/cm^2 for 100 and 150 µm pitch of square arrays and 1160/cm^2 for triangular ones). The patches contain randomly distributed but parallel-oriented cylindrical pores with a diameter D of 180 to 300 nm. For the wire growth, one side of the templates was sputter-coated by an Au layer (~ 20 nm), which was further enhanced by an additional electroplated Cu layer (5-10 µm, *Cupatierbad,* Riedel GmbH). This conductive backing layer with improved mechanical stability served as the substrate of the cathode. The deposition of Au into the pores was performed in a galvanic cell using a potassium dicyanoaurate (I) solution (*Puramet 402,* Doduco GmbH) with 10 g/l of gold at 50°C and a gold rod as anode [Liu06a]. The deposition of Pt was performed in a galvanic cell using an alkaline Pt-solution at 65°C.

The length of the wires and, thus, their AR was controlled by deposition time and by recording the deposited charge. After the dissolution of the template in a CH$_2$Cl$_2$ solution, regular patch arrays of freestanding nanostructures on Cu-Au substrates were achieved as shown in the Fig. 4.22. Finally, all samples were carefully assembled on aluminum cathode holders of about 8 mm in diameter. The main parameters of the fabricated structured Au and Pt-NW cathodes are summarized at the table 4.2.

Table 4.2: Overview of parameters of the structured MNW cathodes: NW site density f, array pitch P, patch diameter Ø, nanowire diameter D, length L and aspect ratio AR.

Cathode	Au-NW cathodes							Pt-NW cathodes		
	square arrays					triangular arrays		square arrays		
	A	B	C	D	E	F	G	Pt182	Pt186	Pt189
f [cm^{-2}]	10^6	10^7		10^8		10^7		10^7		
P [µm]	150	100 150		100		320		150		
Ø [µm]	50	50		50		180		50		
D [nm]	311	270		270		200		170		
L [µm]	25 ±3	8 ±1	28 ±2	7 ±1	28 ±2	11 ±1	14 ±1	4 ±1	12 ±1	25 ±3
AR	161	60	207	52	207	110	140	47	141	294

4.2.2. FE homogeneity, alignment, and current limits

The morphology of typical NW structures before and after the FE measurements was investigated by the SEM (Phillips XL30 at GSI and Phillips XL30S at Univ. Wuppertal). FE sites distribution from the whole NW cathodes (xy-surface tilt-corrected with respect to a flattened conical anode of 160 µm diameter to achieve a constant gap Δz within ±1 µm for the full scan area of 50 mm^2) was non-destructively imaged with the FESM in vacuum of 10^{-9} mbar. After in-situ replacement of the conical by a needle anode with a tip apex radius R_a of 15 µm, the alignment, homogeneity, and efficiency of FE from selected patch arrays were examined and the integral current of arbitrarily chosen metallic NW patches was measured. Sharper W tip anodes ($R_a < 3$ µm) were used to resolve FE from single nanostructures. In comparison to unstructured cathodes, it is remarkable that the array structure enabled us to correlate the FESM and the SEM results for better understanding of the properties of nanostructures.

The macroscopic electric field E was calculated as ratio of applied voltage to an effective distance d between a tip anode and a relevant emitter. This distance d was determined for each FE site by the linear extrapolation of the PID-regulated $V(z)$ curve (for 1 nA fixed FE current, see Chapter 3) to zero voltage, which corresponds to contact between anode and emitter. The effective field enhancement factor β_{eff} of the patches were derived from the measured I-V data after field calibration for each FE site and using the modified FN equation, assuming the work function of 4.9 eV for Au and 5.3 eV for Pt. It should be mentioned that for d values close to or smaller than the mean NW length, the effective field enhancement factor β_{eff} is not only given by the intrinsic field enhancement of the NW, but is also influenced by the d value as suggested by the two region FE model of *Zhong et al.* [Zho02] (see Chapter 3). The influence of the electrode gap geometry including the real shape of the anode or gate on the FE parameters should be noted for a triode device development.

Rather different freestanding nanostructures were obtained in the patch arrays of the Au-NW cathodes as shown in Fig. 4.23. Depending on the number density and AR, the wires are either solitary vertically aligned or they agglomerate into conical clusters. Solitary wires are typical for low number density (10^6 cm^{-2}, cathode A) and for short (< 15 µm) NW at medium (10^7 cm^{-2}, cathodes B, F and G) and high number density (10^8 cm^{-2}, cathode D).

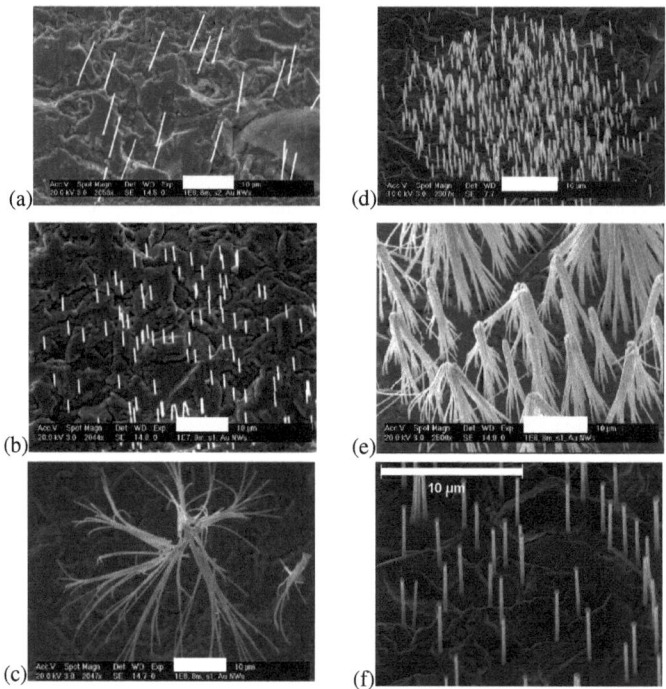

Fig. 4.23: Typical SEM images of single patches of Au-NW of the cathodes (a-e correspond to A-E and f to G cathodes, F is very similar to G). The white scale bars corresponds to 10 μm.

In contrast, most of long NW at medium and high number density is agglomerated into conical clusters probably by bending during the wet dissolution process of the polymer template. Clustering is dominant for NW of high AR (high ductility) and when the average wire distance is much smaller than their length. It is notable that there are only few clusters per patch for medium but many competing clusters for high number density of NW. It is rather puzzling to predict the exact form of clusters, especially in case of few clusters per patch, but some control is still possible. It can be done by adapting NW density and AR. Clustering of NW is favorable as it leads to decreased mutual shielding effect by better separation of emitters, improved current carrying capability by sharing of FE current between many individual NW, and improved heat conduction. It should be mentioned that some of NW are laying flat on the substrate (especially for the cathode E with long dense NW). Most probably this is caused by the wet template dissolution due to insufficient contact strength of NW to the substrate.

Strongly comparable results in terms of morphology were obtained for the structured Pt-NW cathodes. They revealed the same type of nanostructures as Au NW for similar fluence and NW dimensions as shown in the Fig. 4.24.

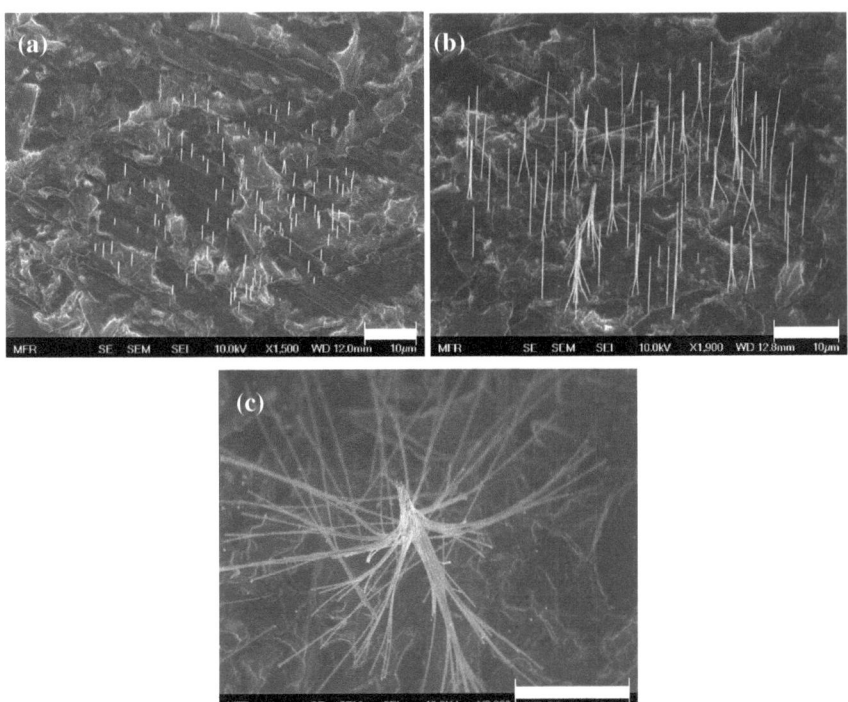

Fig. 4.24: Typical SEM images of single patches of Pt-NW of the cathodes (a, b, c correspond to Pt182, Pt186, Pt189). The white scale bars corresponds to 10 µm.

Low-resolution FESM voltage maps of the whole area of the structured Au-NW cathodes confirm a relatively homogeneous emission from the patch array regions somewhat limited by slight non-planar surface. The resulting map of the cathode with 10^7 NW/cm^2 in Fig. 4.25 [Nav08] clearly shows the separation of four patch areas with two different spacing. Spots in the wire-free bridges are probably due to NW moved during sample handling. Strong emitters appear at around 10 V/µm, whereas weak NW require up to 40 V/µm. Similar low-resolution FE site maps of the other cathodes are shown in Fig. 4.23, where the patch areas exhibited as well.

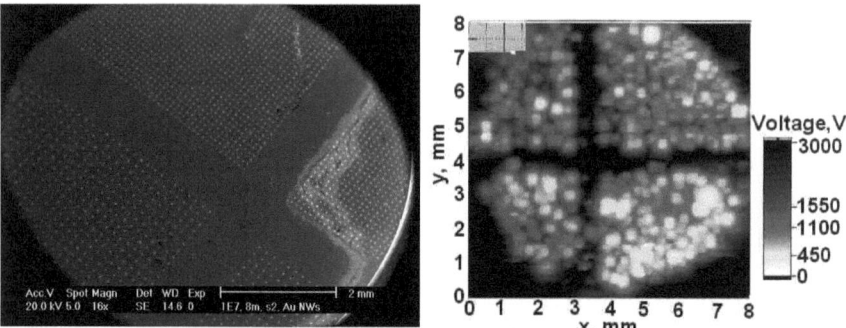

Fig. 4.25: The SEM image of the structured array with 10^7 Au-NW per cm^2 (left) and corresponding FE map obtained by the voltage required for a local current of 1 nA (right, scanned area 64 mm^2, anode Ø 160 μm, electrode gap Δz = 50 μm). Please note that FE map is rotated with respect to the SEM image.

Typical medium resolution FESM maps of about 1 mm^2 scan area of the cathodes shown in the Figs. 4.26 and 4.27 demonstrate that the efficiency and alignment of the FE sites considerably depends on the type of nanostructures in the patches. Solitary NW with low number density provides a moderate FE efficiency with only 32 % of emitting patches at fields up to 50 V/μm (cathode A), whereas about 70-90 % of the patches with medium and high number density emit at 30-40 V/μm. Aligned FE for patch arrays with 100 μm pitch has been found to be more pronounced for short NW due to unordered cluster creation of long NW resulting in a larger scattering of field enhancement factor.

The best result in terms of alignment of square arrays was reproducibly achieved for arrays with $f = 10^7$ cm^{-2} of short NW (see cathode B in Fig. 4.23). Medium resolution FESM maps of patch arrays with 150 μm pitch show good alignment for the cathodes B and C (Figure 4.27), i.e. for both NW length. In contrast, high NW density ($f = 10^8$ cm^{-2}) leads to less alignment because of preferred FE from the patch edges due to enhanced mutual shielding. This means that the position of the FE sites might be rather on the circumference instead of the interior of the patches. In order to get structured NW cathodes with sufficient alignment for triode devices, therefore, the pitch should be at least twice the patch diameter.

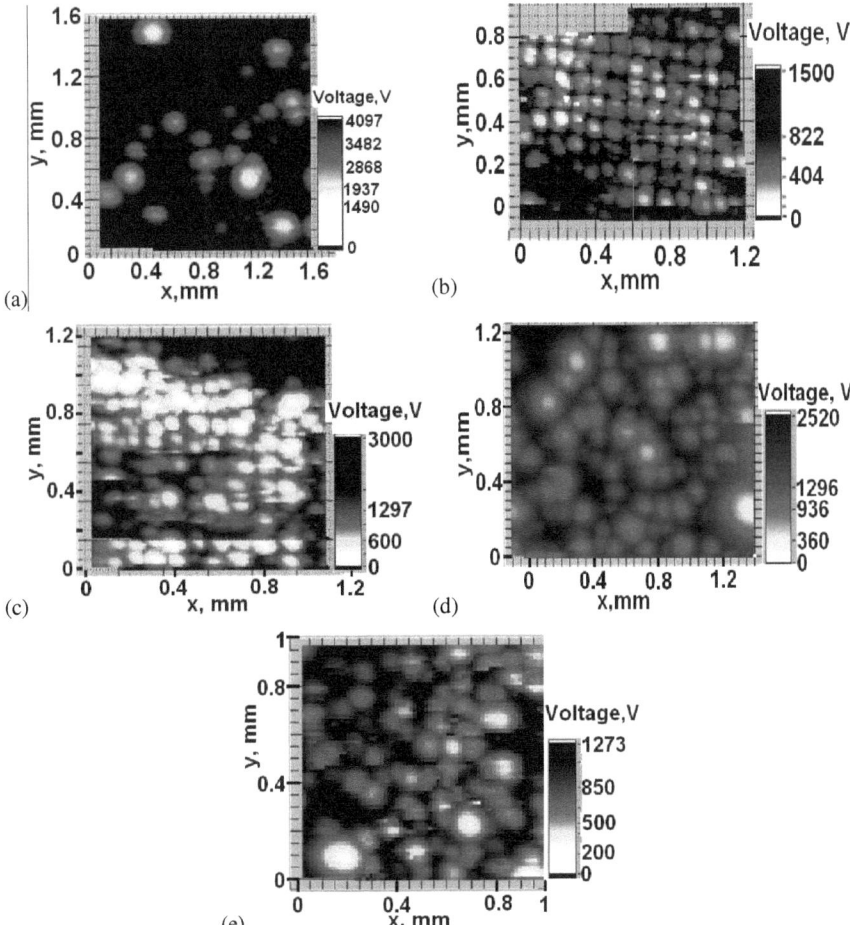

Fig. 4.26: Regulated voltage maps V(x,y) for 1 nA current of cathodes A (10^6 NW/cm^2; a), B, C (10^7 NW/cm^2; b, c) and D, E (10^8 NW/cm^2; d, e) cathodes with short (a, b, d) and long (c, e) NW in narrow spaced arrays (P = 100 μm, scanning anode tip $\varnothing_a$ = 30 μm, d ≈ 20 μm).

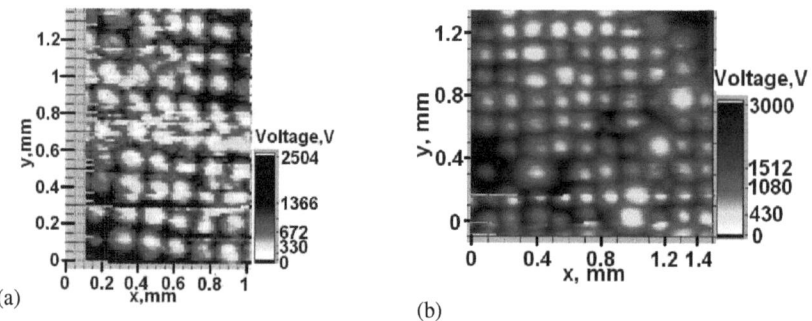

Fig. 4.27: $V(x,y)$ for 1 nA of cathodes B, C (10^7 NW/cm^2) with short (a) and long (b) NW in wide-spaced patch arrays ($P = 150$ μm, $\emptyset_a = 30$ μm, $d \approx 20$ μm).

Further high-resolution FESM studies of individual patches (Fig. 4.28) confirmed the random distribution of FE sites over a patch. They prove the existence of several competing

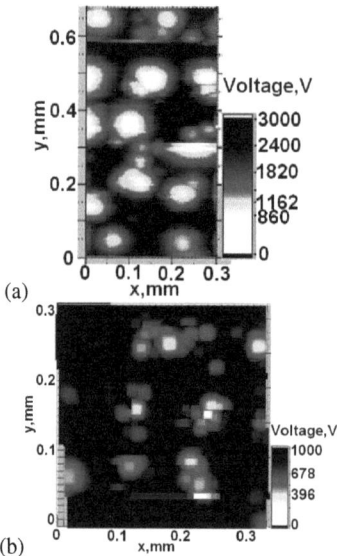

Fig. 4.28: High resolution $V(x,y)$ for 1 nA of patch arrays of short Au-NW: (a) cathode B, $f = 10^7$ cm^{-2}, $P = 150$ μm, $\emptyset_a = 5$ μm; (b) cathode D, $f = 10^8$ cm^{-2}, $P = 100$ μm, $\emptyset_a = 3$ μm.

emitters within a patch, especially for high number density of short NW as expected from the corresponding SEM results. The dominant emitter is not necessarily in the center of a patch and it might cause some emission image in the spacing between the patches. Multiple emitters per patch, however, contribute to the high efficiency and might be important for long-term stability of Au-NW cathodes due to substitution of destructed emitters.

The current-voltage curves and the maximum current I_{max} were locally measured for 10-30 patches per cathode with square arrays of Au-NW. In general, strong processing effects were observed especially during the first voltage rise resulting in a stabilized FE from the

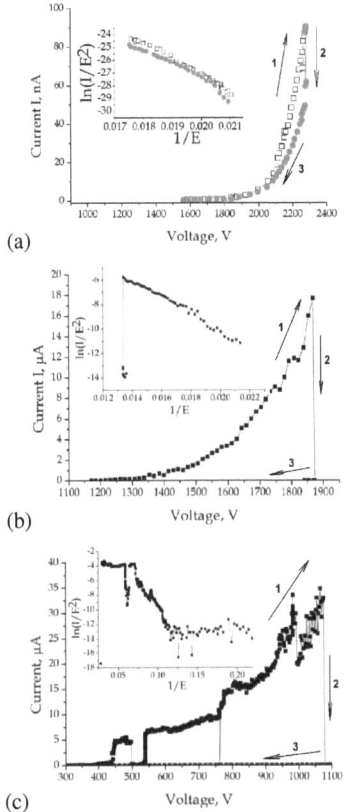

Fig. 4.29: Typical I-V curves and FN plots (insets) of single patches with long Au-NW for $f = 10^6$ cm^{-2} of cathode A (a), 10^7 cm^{-2} of cathode C (b) and 10^8 cm^{-2} of cathode E (c) measured with $\varnothing_a = 30$ μm at $d \approx 20$ μm. Please note the different current scales. Arrows indicate increase of voltage (1), current drop due to full or partial destruction of a FE site (2) and decrease of voltage (3).

patches after few minutes. Possibly, it is the result of the adsorbates effects, nanostructural changes of NW tips and/or movement of loose NW in electric field. Patches with long solitary NW showed some current jumps and finally straight FN behavior (Fig. 4.29(a)) up to moderate levels of I_{max} < 450 nA. In contrast, 40 % of the patches with few NW clusters yielded stable I-V curves up to µA current levels (Fig. 4.29(b)). Most of the patches with a large number of NW clusters, however, carried stable currents only up to 50 nA, and only a few up to 40 µA with big current jumps (Fig. 4.29(c)). These results indicate current limitation or stepwise destruction of solitary or clustered NW [Dan07b]. According to the electro-thermal model calculations [Nav08, Nav09a, Nav10d], performed with Comsol Multiphysics [Com], resistive heating of Au-NW at such current levels can be neglected even for significantly reduced electrical and thermal conductivities as compared to the bulk values [Kar07]. A significant uniform heating (ΔT > 200 K) is obtained, however, with assumption of very high contact resistance and/or a strongly reduced contact area (Fig. 4.30). The contact constrains are also suggested from the SEM images revealing ring-like contact area [Nav09a, Nav10d] of NW to the substrate (see Fig. 4.37 below).

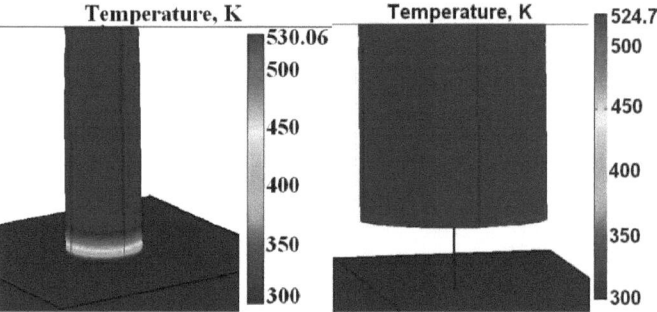

Fig. 4.30: Temperature distribution along Au-NW with high contact resistance to the substrate (left), and with a strongly reduced contact area (right) resulting from electro-thermal model calculations [Nav09].

Despite of some remaining instabilities of I-V curves, finally the local electric onset field E_{on} for a current of 1 nA and effective field enhancement factor β_{eff} of the patches were derived from the measured data. The lowest E_{on} of 4 V/µm was obtained for a patch with well-separated long NW (cathode A), but the highest β_{eff} of 630 resulted for a patch with many NW clusters (cathode E). The hyperbolic correlation between the E_{on} and β_{eff} values expected for constant emission area was more pronounced for solitary than for clustered NW [Nav08]. The mean value of β_{eff} and I_{max} determined for all type of patches scatters significantly and seems to be independent of the aspect ratio as shown in the Fig. 4.31. The smallest spread of both

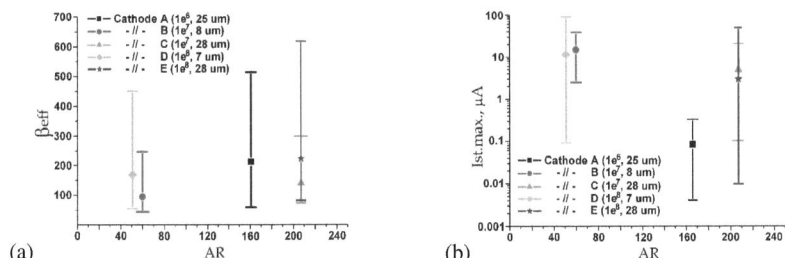

Fig. 4.31: Mean value (dots) and range (bars) of measured field enhancement factor β_{eff} and maximum stable emission current I_{max} for all patch types vs. aspect ratio of Au-NW.

figures of merit is achieved for short Au-NW of medium number density, while the largest spread results for many competing NW clusters. The correlation between the local β_{eff} and I_{max} values found for random unstructured Au-NW ensembles with $AR = 110$ [Dan08] could not be confirmed for the NW patches. These results reflect the complex interplay of the shape, number density and cluster formation of NW on the FE homogeneity and current carrying capability of Au-NW patches. The mutual shielding effects within the NW patches have a strong influence on their field enhancement as well known for densely grown CNT [Nil00].

Sufficiently comparable results in terms of homogeneity and alignment of emission were obtained also for the structured Pt-NW cathodes with square arrays as shown in the Fig. 4.32. According to the smaller patches of the quadratic array, less Pt-NW per patch are obvious in the SEM image thus resulting in slightly reduced FE efficiency and alignment of the patches. Single Pt-NW patches provided nearly the same average electric onset field (28 V/μm for 1 nA) as Au-NW patches. The derived mean effective field enhancement factor (189 for the work function of 5.3 eV) was slightly higher, possibly indicating sharper edges of the actual Pt emitters. The maximum stable current of the individual Pt-NW patches varied again between 1 and 50 μA, despite of the higher melting temperature of Pt (2042 K) as compared to Au (1336 K). Therefore, current limits of the nanostructure emitters of both noble metals seem to be caused by some other parameters, which are not under control yet.

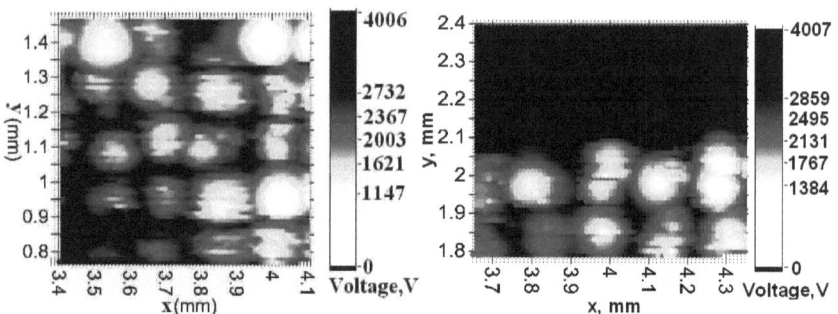

Fig. 4.32: V(x,y) for 1 nA of cathode Pt186 (f = $10^7/cm^2$, P = 150 µm, L = 12 µm, $\emptyset_a$ = 8 µm, d ≈ 30 µm).

Structured Au-NW cathodes with triangular arrays, produced for matched gate for planned triode investigations, displayed a high emission efficiency of the patches up to 100 % [Nav09b, Nav10d] in the FESM in diode configuration. However, the alignment of the patch emitters was limited by random NW position and better for the structures with larger pitch to patch ratio.

Figure 4.33 shows an optimized cathode result, i.e. all patches emit well aligned at electric fields around 25 V/µm. The irradiation fluence of $10^7/cm^2$ and NW length of 11 µm of the cathode F yields about 4-8 emitters per patch for this sample. The integrally measured I-V curves and current carrying capability of the single patches varied strongly. A mean effective

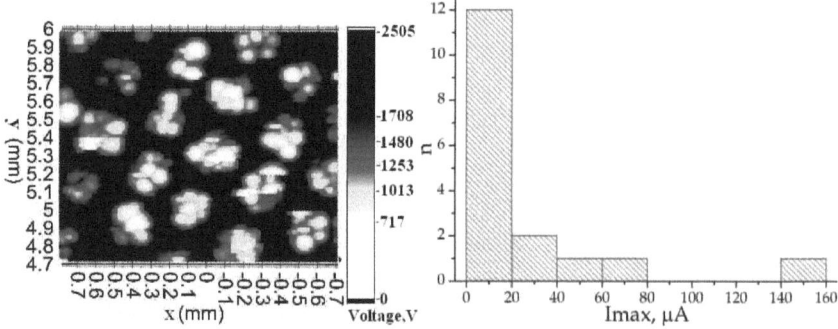

Fig. 4.33: Regulated voltage map for 1 nA current (scanned area 2 mm^2, anode Ø 30 µm, gap 40 µm) of triangularly structured Au-NW ($10^7/cm2$, length 11 µm, Ø 200 nm) and histogram of maximum current of individual patches (anode Ø 100 µm).

field enhancement factor of 122 was derived from the FN slopes of 14 patches. This value is close to the aspect ratio of NW and expected for a gap size in the order of NW length [Nav08]. In contrast to rather reproducible electric onset field, sufficiently different values for the maximum current of 17 patches were also obtained as shown in the Fig. 4.33. Most emitters were partially destructed starting at few µA, and the best patch provided a stable current of 140 µA at 45 V/µm. Stable FE currents of 100 µA per patch, i.e. current densities above 100 mA/cm^2 would make such NW cathodes suitable for power devices.

Several integral tests of structured Au-NW cathode with triangular patch array were performed in adjustable diode and triode configuration using a luminescent screen as anode and a matched gate for the triode. In pulse diode mode, (2 ms pulse length, 10 % duty cycle), the maximum current density of 0.5 mA/cm^2 at about 10 V/µm was limited by strong current fluctuations and partial phosphor evaporation. First best triode results showed a moderate dc current density j < 0.5 mA/cm^2 at 11.8 V/µm and a low ratio of anode to cathode current of Ia/Ic ~ 37% limited by the difficult gate matching. Figure 4.34 shows relations between cathode, gate and anode currents at increasing field between gate and cathode; and luminescence screen (anode) image at maximum achieved cathode current.

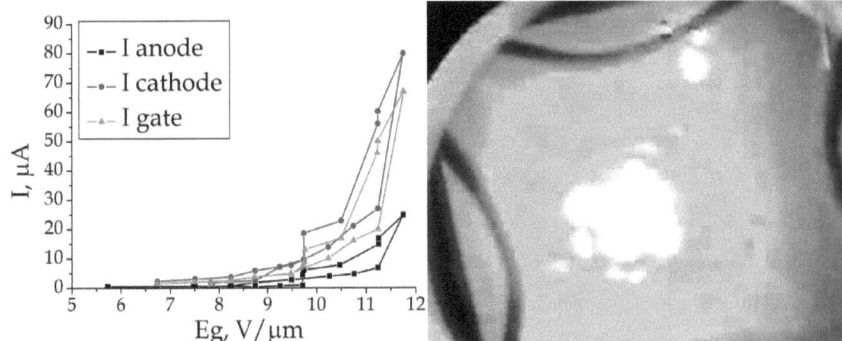

Fig. 4.34: Relations between cathode, gate and anode current at increasing gate-cathode field and fixed anode voltage of 6.5 kV (13.5 mm gate-anode spacing) (left); emission image acquired at E = 11.8 V/µm and dc cathode current I_c = 80 µA. Please note bright background due to high reflectivity of gold.

4.2.3. FESM-SEM correlation study

Finally, we have started to look for correlations between the measured I-V curves and resulting SEM images of selected Au-NW patches in order to understand the origin of current limitation of NW. In the Fig. 4.35, examples of processed patches showing typically observed

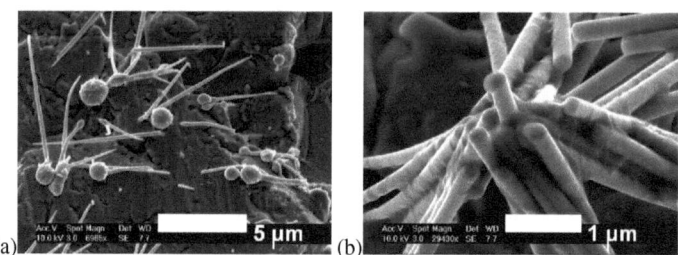

Fig. 4.35: Typical SEM images of selected Au-NW patches ($f = 10^8$ cm^{-2}, L = 7 µm): destruction of single NW at I ≈ 3 µA (a) and NW clusters at I ≈ 90 µA (b).

destructions are compared for two types of nanostructures. The tips of individual NW were obviously destroyed by successive melting which results in spherically shaped tips and probably explains rather moderate emission currents. In contrast, NW clusters provide a better mechanical stability and a higher maximum current because of the mutual support and only partial melting of NW tips. It is most remarkable that many of cluster-forming NW in the Fig. 4.35(b) reveal some ripples on their surface. It might reflect thermally driven morphological changes much below the bulk melting temperature (the Rayleigh instability), thus proving their contribution to total current of the cluster. With respect to the electro-thermal model calculations mentioned above, these melting features indicate rather large contact resistance between NW and the substrate. Unfortunately, the contact resistance cannot be measured directly because it is problematic to ensure good electrical contact between a wire and a tip anode as a probe without mechanical destruction of the wires. Moreover, in-situ high-resolution SEM control would be required, which cannot be performed within the FESM.

Therefore, we have started to look for systematic correlations between the morphology and the FE properties of selected Au-NW patches. A series of high resolution SEM (JEOL JSM-7401) images of their initial and final morphology were taken before and after the local FE measurements. As shown in the Fig. 4.36, the virgin patches contain randomly distributed but mostly solitary vertical-aligned NW. After processing at current limit (~1 μA), about 50 % of the patch is destructed due to partial melting and disruption of NW.

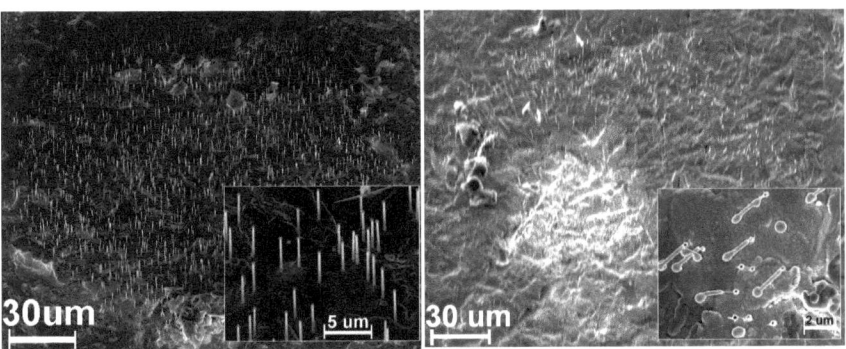

Fig. 4.36: Typical SEM images with high resolution insets of the same Au-NW patch before (left) and after current processing (right, rotated 90°) where the bright area reflects Au coating due to NW destruction.

In case of mechanically disrupted NW, ring-like footprints on the surface of the cathodes can be seen (Fig. 4.37(left)). Moreover, NW tilted by FE current flow reveal a more tube-like structure of the contact area (Fig. 4.37(right)). Often, the remaining NW in the processed patches show partial deformations, especially bubbles which start to grow mostly from the top (see Fig. 4.36). Therefore, the melting seems to start at much reduced temperatures [Kar06]. In contrast, ripples already occur on the surface of virgin NW most probably due to the roughness of the pores resulting from the ion tracks.

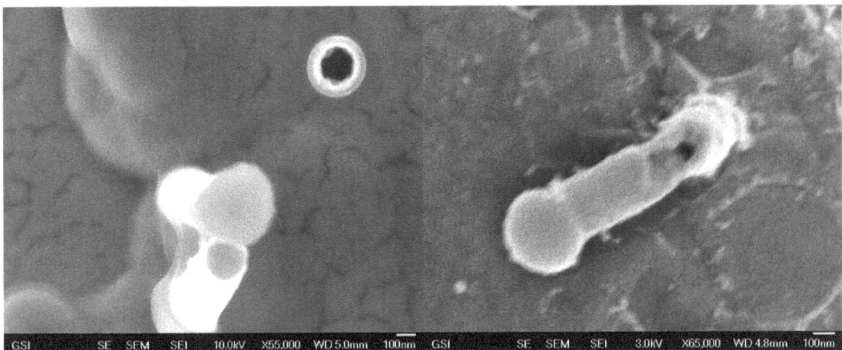

Fig. 4.37: SEM images showing a ring-like contact area as footprint of a mechanically disrupted NW (left) and with a partially molten NW revealing a more tube-like structure of the real contact (right).

Geometrical constrictions of the contact region between Au-NW and the copper substrate has led to a poor conductivity and cooling of NW. Therefore, nearly uniform heating of NW and finally melting results already at current values in the nA range. Further optimization of NW patch arrays for vacuum nano-electronic devices will require appropriate modifications of NW fabrication (crystallinity and geometry) and cathode preparation techniques.

In conclusion, systematic FE measurements for the various types of structured Au and Pt-NW cathodes have given valuable hints to optimize regular patch arrays of random nanostructures for triode applications. The best result in terms of efficiency and alignment of the FE sites was achieved for double-spaced patches of 50 µm diameter with Au-NW at number densities of 10^7 cm^{-2}. Multiple nanostructures per patch contribute to the high emission efficiency of the cathodes. High field enhancement and low onset fields are fairly well correlated for solitary NW, but achievable maximum currents vary strongly for both types of nanostructures. First correlation studies between the FESM maps and the SEM images of selected patches revealed the partial melting of solitary and cluster NW as main cause for obtained current limits. Thermal model calculations, SEM-FESM-SEM correlation studies and comparison of the results of Au and Pt NW patches revealed contact problems as main source of NW heating and destruction. These results have given suitable hints for a modification of the fabrication steps to improve the FE homogeneity, current strength and stability of structured metallic NW cathodes.

5. Parasitic FE from Nb surfaces

EFE from particulate contaminations or surface irregularities is one of the main field limitations of high gradient superconducting niobium cavities [Dan07a] required for the XFEL and the international linear collider (ILC) [Ilc]. While the number density and size of particulates on metal surfaces can be seriously reduced by high pressure ultra pure water rinsing (HPR) [Kne95, Kne96], dry ice cleaning (DIC) [Pro01, Res04, Res05, Res07a, She90], and clean room assembly of the accelerator modules [Res05a, Res05b, Kno99a], the influence of surface defects of electropolished (EP) [Sai03, Mat05, Kne90] and electron-beam-welded (EBW) Nb surfaces on EFE has been less studied yet. Therefore, systematic measurements of the average and local surface roughness of typically prepared Nb samples, some of which were cut out of a real nine-cell polycrystalline Nb cavity have been performed. By means of OP combined with AFM, large scanning ranges as well as high-resolution zooms into found defect areas were obtained. OP is considered as a fast tool for the quality control of the final surface preparation of Nb accelerating cavities. Electric field enhancement factors were derived from the height and sharpness of the localized defects. After HPR of selected Nb samples at DESY, correlated FESM and high-resolution SEM investigations have been done in order to reveal the influence of the defect geometry on the expected EFE and accordingly on the performances of the accelerator modules.

5.1. Nb surface preparation
5.1.1. Overview of the cavity fabrication methods

Fabrication of the accelerating superconducting radio frequency (SRF) cavities is a complex scientific and engineering procedure containing many steps. The cavity fabrication starts from Nb sheet metal forming and passes through deep drawing, spinning or hydroforming of half-cells, trimming, electron beam welding etc. and contains lots of intermediate purification, cleaning and inspection steps [Sch05, Wen00, Sch81, Ade09]. The key specifications of the sheet Nb are residual resistivity ratio (RRR > 300) and impurity concentration, for example, Ta content should be less than 500 ppm [Sin03, Kne05b, Pad09]. Many of the fabrication steps are conjugated with pollution, damage, and mechanical stress of the surface. The surface damage layer has been established to be 100 - 200 µm deep [Kne97] so that a bulk chemistry to remove this amount of material is necessary. In this step, the vapor from welding deposited on the inner surface (~ 25 µm) is also to be removed to arrive at a clean surface. If the welds have imperfections, tumbling or centrifugal barrel polishing (CBP)

[Hig01] can be used for smoothing the surface, following by a buffered chemical polishing (BCP) [Lil04a], which removes the tumbling abrasive embedded in the surface (usually about 50μm). BCP is used for the heavy etch for many applications, but the EP is needed to provide a smooth surface for those applications demanding surface field above 80 mT, corresponding roughly to E_{acc} > 20 MV/m. Both BCP and EP procedures carry a risk of hydrogen absorption so that furnace treatment is necessary to ensure a hydrogen-free cavity [Pad09].

The final chemical treatment is a light etch (about 20-30 μm material removal) either by BCP or EP depending on the target field level. If done properly, there is little risk of H contamination. After thoroughly rinsing the acid residues with high-purity water, the cavity is transported to a class 10-class 100 clean room, where the inside surface is exposed to the HPR (100 bar) with high-purity water jets for many hours. The main goal is to scrub the chemical residues and particulate contaminants, which may cause the FE or thermal breakdown (quenche) of the cavities. After HPR, the cavity dries in the clean room for few days. To remove the high-field Q-drop [Pad09] and reach the highest field, an electropolished cavity needs to be baked at 120°C for 48 h.

The DIC has emerged recently to be very effective tool for final surface cleaning. The best DIC single crystal Nb sample did not provide any FE up to the electric surface field of 250 MV/m [Dan07a]. Further, the removal of field emitting particulates down to 400 nm size and partial smoothing of edges of the protrusions by the DIC of Nb surface was also reported.

Cleanliness is of great importance during the surface preparation stages. All cleaning and assembly is carried out with high-purity solvents/water in clean rooms [Kno99a]. Much attention to the surface preparation and cleanliness techniques has suppressed significantly the EFE from the cavity surface and thus improved regular cavity performance at high accelerating gradients, e.g. up to about E_{acc} = 30 MV/m for nine-cell 1.3 GHz structures [Cio06].

An approach towards improving the cavity fabrication for future linear accelerators like XFEL and ILC has been made using the BCP treated large grain Nb (LGNb) [Dan09] or single crystal Nb (SCNb) instead of EP polycrystalline Nb. They might be less expensive due to the elimination of sheet fabrication and related processes. Preliminary tests of single cell cavities made from LGNb have yielded E_{acc} up to 45 MV/m, which is one of the highest value achieved yet [Kne05a]. Further research steps on the multi-cell structures made from LGNb or SCNb are required before they can replace polycrystalline Nb. It has also been reported recently that the grain boundaries on LGNb cavities provide some, although not dominant, contribution to the hot spots in corresponding thermal maps. Since the grain boundaries get easily contaminated by the segregation of impurities during usual bakeout of cavities, it is of interest to investigate their role for FE as well.

5.1.2. Samples for the surface study

Altogether six flat and five curved Nb samples were tested [Nav09d]. The flat ones with Ø 26.5 mm were machined from the high-purity Nb sheets (standard for the SRF cavity production is RRR > 300) and welded to support rods (Fig. 5.1). The samples are specially made for a quality control of chemical treatments of the cavities [Mül98]. The support rods hold the samples in the main coupler port of nine-cell cavities during the chemical treatments, as well as in the OP, FESM, and SEM setups. Two special marks on the circumference have been made to re-identify their angular position during the surface studies. The curved samples were cut out of the nine-cell cavity, which suffered from strong quenching at moderate accelerating field of 16 MV/m. These samples were taken from different interesting surface areas: two (one) from the flat part of the lower (upper) half-cell, one from the iris, and one from the equator weld regions. All samples have obtained comparable EP and BCP etching as well as HPR but the flat ones were not kept in clean and mechanically safe conditions. As a result, the surface contains some artificial scratches, which were used for the correlation study between the geometry of the defects and their FE properties. As a preparation step for the first surface study, the flat samples have been cleaned in an ultrasonic isopropyl alcohol bath and dried with an ionizing gun using pressurized filtered nitrogen (5.0). After that, the samples were again degreased by washing, cleaned in the ultrasonic water bath and finally cleaned by means of the HPR as described below, dried with the ionizing gun, closed with caps, and sealed in plastic bags. The caps (Fig. 5.1) are made of aluminum and serve for a mechanically safe and dust-free transport of the samples. To avoid any contaminations, the caps were also degreased by washing, cleaned in the ultrasonic water bath and finally dried as described above, but have not been treated by the HPR. The HPR as well as preparation and packing of the samples have been done in the same conditions as for the cavities, i.e. in a clean room with a horizontal laminar airflow. During the FESM investigations, the caps were removed for the first time only in the preparation chamber of the FESM under vacuum of at least 10^{-7} mbar. Extraction of the samples from the FESM and transport to the SEM was done with the same caps, which were stored and installed back in the preparation chamber. During the SEM

Fig. 5.1: The flat Nb sample with the protection cap.

investigations, the caps were removed after fixation of the sample and shortly before closing of the SEM chamber. Handling of the samples after the HPR was done in a way eliminating contaminations and reducing damage of the sample as much as possible.

All the samples reported here were got from A. Matheisen, X. Singer and D. Reschke from DESY and they had been already chemically treated, while the HPR and partial surface cleaning were done by D. Reschke with my participation.

5.1.3. Electropolishing (EP) of Nb

EP is an etching method in which material is removed by acid mixture under flow of electric current. Sharp edges and burrs are smoothed out and a very glossy surface can be obtained. The electric field is high at protrusions, so these will be dissolved first. On the other hand, the field is low in the grain boundaries and little material will be removed here while in the BCP process strong etching is observed in the boundaries between grains. EP of niobium cavities has been known for 40 years. The most widely used electrolyte is a mixture of concentrated hydrofluoric HF and sulfuric H_2SO_4 acids in volume ratio of 1:9 [Die71, Res05a], which was also used for preparation of the samples described here. A pulse electric current was used in a horizontal EP setup for superconducting niobium cavities prepared at CERN in collaboration with Karlsruhe in 1979 [Cit79]. A continuous method for horizontal EP has been developed at KEK in 1989 [Sai89]. The chemical processes are as follows [Kne80, Pon86]:

$$2Nb+5SO_4^{2-}+5H_2O \rightarrow Nb_2O_5+10H^++5SO_4^{2-}+10e^-,$$
$$Nb_2O_5+6HF \rightarrow H_2NbOF_5+NbO_2F \cdot 0.5H_2O+1.5H_2O,$$
$$NbO_2F \cdot 0.5H_2O+4HF \rightarrow H_2NbF_5+1.5H_2O.$$

Most systems for the EP of accelerator cavities are horizontal as shown in the Fig. 5.2.

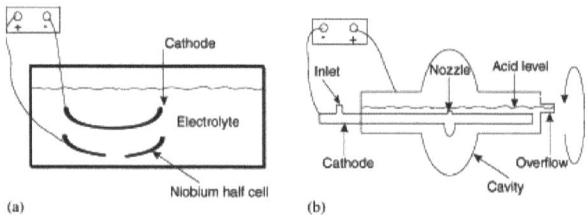

Fig. 5.2: Illustration of a half-cell EP system (a) and of a cavity EP system (b).

The advantage is that produced gases (mainly hydrogen) are rapidly removed from wetted niobium surface [Cit79]. Gas bubbles sticking on the surface could lead to etching pits. In a vertical setup, these bubbles would move slowly upwards and create axial wells. A drawback of the horizontal arrangement is that the cavity has to be rotated. There is some difficulty to achieve a leak tight rotary sleeve for the acid mixture. In addition, the removal rate is reduced by a factor of two since the surface is immersed only half of the time in the acid to allow the hydrogen gas to escape through the upper part of the beam tube. EP of the samples was done at DESY using their horizontal EP system during preparation of the real Nb cavities named AC114 and AC123 in case of the flat samples. The curved ones are originated from the Z111 cavity. All samples have obtained comparable EP up to 144 µm.

5.1.4. Buffered chemical polishing (BCP) of Nb

Niobium metal has a natural Nb_2O_5 layer with a thickness of about 5 nm which is chemically rather inert and can be dissolved only with the hydrofluoric acid HF. The damage layer has to be removed in order to obtain a surface with excellent superconducting qualities. One possibility is chemical etching, which consists of two alternating processes: dissolution of the Nb_2O_5 layer by HF and re-oxidation of the niobium by a strongly oxidizing acid such as nitric acid (HNO_3) [Lil04b, Gme70]. To reduce the etching speed a buffer substance is added, for example phosphoric acid H_3PO_4, and the mixture is cooled below 15°C. The standard procedure with the removal rate of about 1 µm/min (comparing to 30 µm/min without buffer) is called BCP with an acid mixture consisting of HF(40%):HNO_3(65%):H_3PO_4(85%) in volume ratio 1:1:2. At Tesla test facility (TTF) at DESY, a closed-circuit chemistry system is used in which the acid is pumped from a storage tank through a cooling system and filter into the cavity and then back to the storage. The gases produced are not released into the environment without prior neutralization and cleaning. The cavities are rinsed with low-pressure ultra pure water immediately after the chemical treatment. This final rinse is done in a closed loop until the specific electrical resistivity of the water has reached 18 MΩcm.

By now, compelling evidence exists that the BCP process limits the attainable accelerating fields of multi-cell cavities to about 30 MV/m even if niobium of excellent thermal conductivity is used [Lil04b]. The etching process is accompanied with undesirable effects such as migration of hydrogen into the bulk niobium and strong grain boundary etching. The first effect can be reduced by cooling the acid below 15°C [Röt93]. In principle, grain boundary etching could be suppressed by using the acid mixture with a high etching rate. This, however, would not be advisable in the nine-cell TESLA cavity with an inner surface of about one m^2 since the large amount of produced heat would speed up the reaction even more and preclude a well-controlled etching process.

The roughness of electropolished niobium surfaces is less than 0.1 µm [Ant99] while chemically etched surfaces are at least one order of magnitude rougher. The main advantage of EP is the far better smoothening of the ridges at grain boundaries. The EP of at least 100 µm is needed for both a surface smoothening and a damage layer removal.

All samples reported here have obtained comparable final BCP etching up to 10 µm at DESY in the same cavities as mentioned in the previous chapter.

5.1.5. High pressure ultra pure water rinsing (HPR)

Micro-particle contamination has been identified to be the leading cause of FE. This stresses the importance of cleanliness in all final treatment and assembly procedures. Rinsing with ultra-pure water HPR has found to be the most effective tool for removal of micro particulates and therefore reduction of FE. The HPR has also been effective in reducing FE, which cannot be processed during rf tests [Pad09]. The HPR must be carried out in a class 10–class 100 clean room to prevent re-contamination with the dust. For best cleaning, it is important to avoid drying between final water rinse after chemistry and the start of the first HPR.

The technology of ultra pure water (UPW) processing, handling and quality monitoring is well established due to needs of electronic and semiconductor industry [Gai02]. The experience of the last decade shows, that the application of the water quality between the 0.5 µm Technology ("16 MB integration") and more challenging 0.25 µm Technology ("256 MB integration") [Ast07] has the ability of excellent cavity performances.

It is general practice to apply ultra pure water (> 18 MΩcm) HPR at 100 bar to the niobium cavities as the final cleaning procedure after chemical surface treatments. The methods of how these rinsing steps are done vary from one laboratory to another. The variations/differences of the exact layout of HPR systems are conjugated with system design, pumping system, spray nozzle design, water quality/monitoring, HPR procedure, etc. and depend on technical, administrative and safety regulations.

There is no clear understanding of the force needed at the cavity surface to dislodge residual contamination from chemical processing or handling. This depends on many parameters [Kne95] as an example - the particle size. One has to know the nature of the particles clinging to the surfaces in order to apply an optimal/effective high pressure water jet.

It is important to keep in mind that HPR procedures are only one step in generating contamination-free surfaces; re-contamination can occur during the drying and assembly processes, from contaminated auxiliary parts attached to the cavities, from vacuum systems and/or test stands. At DESY the following procedures are established: after the last EP, HPR is used for 90 minutes in a 9-cell cavity; the nozzle is moved up once, and down again. Each cell is rinsed twice during five minutes. After rinsing, the 9-cell cavity is drying under laminar flow conditions in the class 10 clean room. Assembly of antennas etc. follows. The last step in cavity preparation is 6 times 90 minutes under the conditions mentioned above [Ttc08]. An overview of one of the HPR systems installed at DESY is illustrated in the Fig. 5.3. This system was used for the cleaning of the samples reported here. The samples (two at once) were fixed in a special holder simulating the cavity and got few minutes of the HPR in two cycles as for the cavity cells.

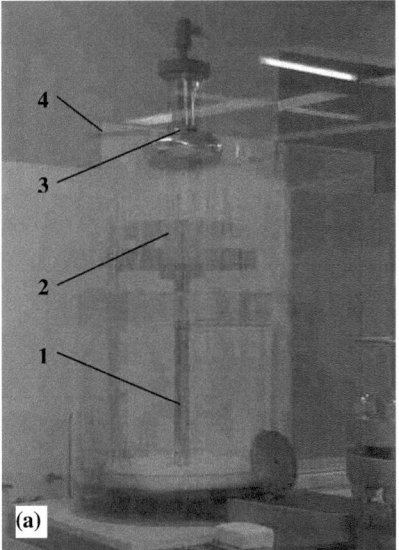
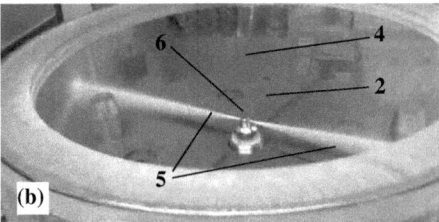

Fig. 5.3: (a) The HPR system for 9-cell cavities at DESY: rotating high pressure water pipe (1), vertically moving support plate (2), 1-cell cavity under rinse (3), protection envelope (4); (b) high-pressure water jet streams (5) emerging from the nozzle (6) [Pad09, Res05a, Res07b].

5.2. Surface roughness and electric field enhancement (EFE)

For measurements of the surface roughness, localization and characterization of the defects the MicroProf® OP was used. The OP measurements of flat samples have been performed in three steps. At first full scans of the sample (27.5×27.5 mm^2) with a lateral resolution of at least 30 µm were made to localize the major defects with respect to the marks at the circumference as shown in the Fig. 5.4.

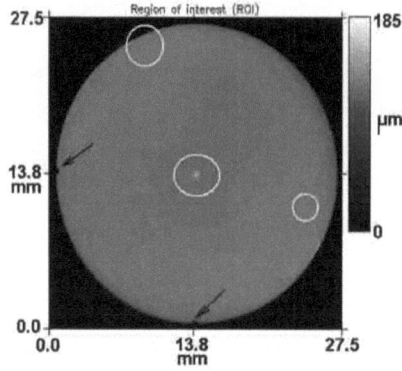

Fig. 5.4. Full profile of the flat Nb sample (30 µm lateral resolution) showing positions of two marks (arrows) and some big defects on the surface (circles).

Then selected regions of typically 5×5 mm^2 size with large defects, which were also clearly seen due to distorted light reflection, were further investigated with at least 10 µm resolution. Finally, interesting defect areas were scanned with maximum resolution of 2 µm over 0.5×0.5 mm^2. It is noteworthy that the full scans of the curved samples were made in several consecutive layers because of the limited height range of the optical profilometer (230 µm). An example of the resulting synthesized 3D view of the equator weld region is presented in the Fig. 5.5. Because of the limited lateral resolution of the optical profilometer, the sharpest features of some localized surface defects were measured with the AFM.

Based on the optical profilometer measurements the average and quadratic surface roughness have been calculated as

$$R_a = \frac{1}{n \cdot m} \sum_{i=1}^{n} \sum_{j=1}^{m} \left(z(x_i y_j) - \bar{z} \right);$$

$$R_q = \sqrt{\frac{1}{n \cdot m} \sum_{i=1}^{n} \sum_{j=1}^{m} \left(z(x_i y_j) - \bar{z} \right)^2},$$

where $z(x_i y_j)$ is the actual and $\bar{z}$ the average value of the profile height and n and m are the number of points in x and y directions. For curved surfaces, these roughness values can also be determined with respect to the nominal surface shape.

Electric field enhancement factor β of defects was estimated from their height h and curvature radius r of upper edges, as described in Chapter 2.

One of the most unusual features that were found on electropolished Nb nine-cell cavities are visible pits. They appear more densely distributed on the lower than on the upper half-cell of vertically treated cavities. We have found such pits on all five curved Nb samples (e.g. in Fig. 5.5) but less on the weld seams. The detailed optical profiles shown in the Fig. 5.6 reveal their size up to 800 µm in diameter and crater-like centers (~ Ø 200 µm) with sharp rims of 5-10 µm height, which might lead to quenches or EFE. Estimation of the maximum electric field enhancement factor leads to $\beta_{E,max}$ values of about 10 for the rims. The origin of the pits and craters is not clarified yet but we consider them as an etching effect (gas bubbles sticking on the surface) due to the finite time required for washing off the acid solution after electropolishing.

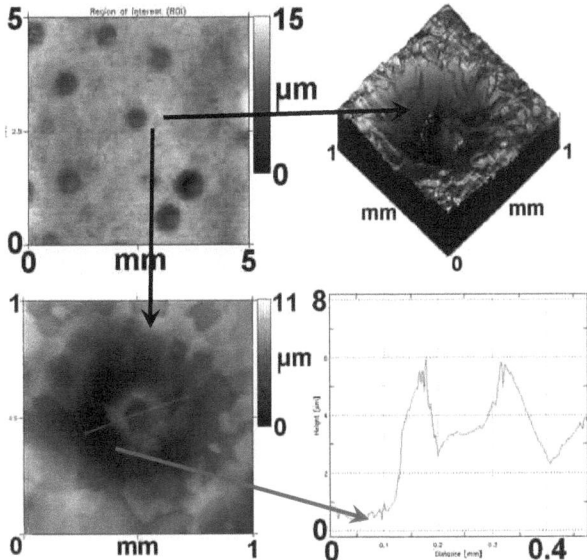

Fig. 5.5: Optical profiles of the upper half-cell part of a tested Nb cavity showing many pits with crater-like elevations and sharp rims in the center. Please note the much enhanced height scale in the 3D view and line profile of the pit.

We have also compared the surface roughness of the curved Nb samples in the half-cell and weld regions. The typical profiles given in the Fig. 5.6 show the surprising result that the weld seams are smoother (R_a = 0.115 μm, R_q = 0.159 μm) than the other regions (R_a = 0.180 μm, R_q = 0.250 μm) probably due to the enlarged grain size (640 μm vs. 152 μm, see Table 5.1). The grain boundaries provide, however, quite similar step height (< 2 μm) and edge radius (< 0.5 μm) resulting in $\beta_{E,max}$ ~ 4-5 which should be negligible for EFE.

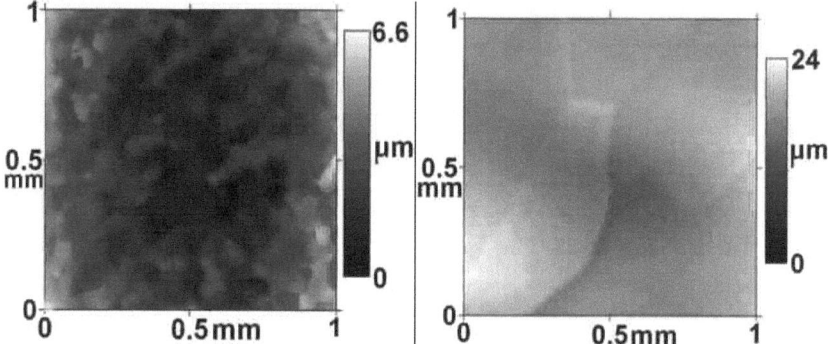

Fig. 5.6: Comparison of high resolution optical profiles of a half-cell (left) and an iris weld (right) regions of the tested Nb cavities.

Table 5.1: Parameters of the grain boundaries in different regions

Half-cell region	Iris-weld region
average grain size ≈ 152 μm	average grain size ≈ 640 μm
step height < 1.964 μm	step height < 1.917 μm
r ≈ 0.45 μm → $\beta_{E,max}$ = 4	r ≈ 0.41 μm → $\beta_{E,max}$ = 4.67
R_a = 0.180 μm	R_a = 0.115 μm
R_q = 0.250 μm	R_q = 0.159 μm

The optical profiles of the flat Nb samples have shown mainly four types of surface defects: particles (rather than foreign material inclusions), scratch-like protrusions, grain boundaries and round hills and holes (see Figs. 5.7, 5.8). The geometrical parameters of these defect categories are summarized in Table 5.2.

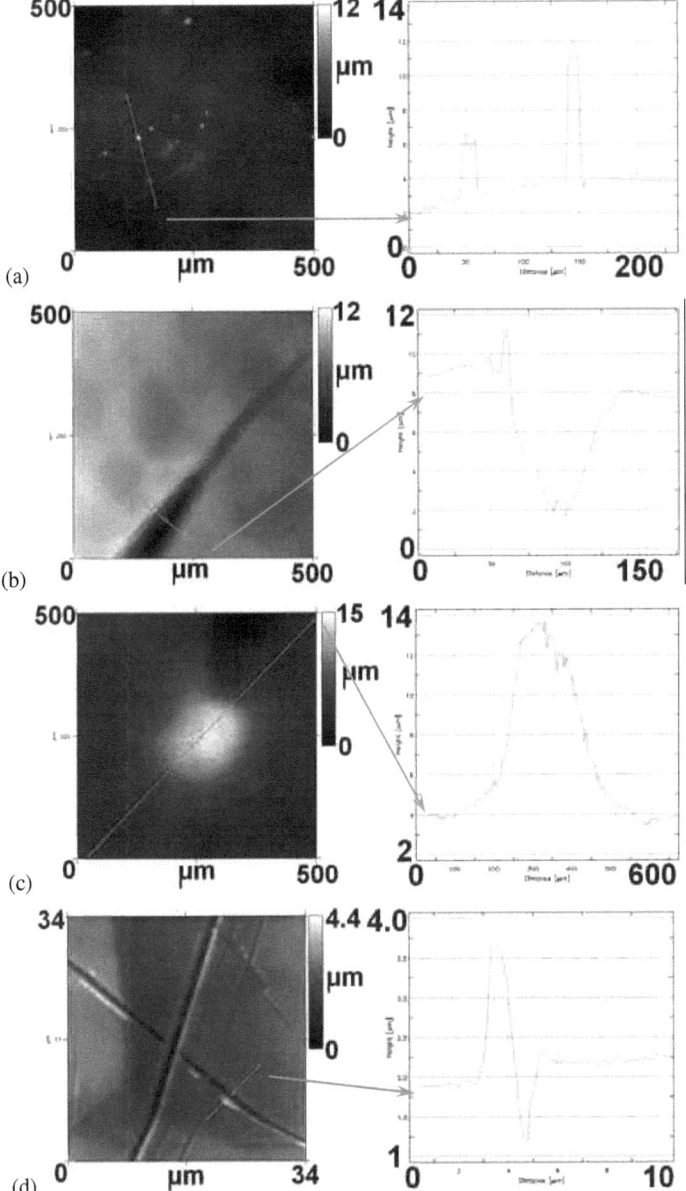

Fig. 5.7: Optical profiles of typical defects found on the flat Nb samples: particles (a), scratches (b), round hills and holes (c) and AFM image of a scratch (d) with corresponding line profiles.

Table 5.2: Geometrical parameters of the defects

Particles		Scratches	Grain boundaries	Round hills and holes
< 5 μm	43 %	4 -100 μm width		
5 - 15 μm	48.4 %	11 μm – 2.7 mm length (on average 326 μm) ridge height < 10 μm	step height < 1.55 μm edge radius < 0.78 μm	height < 17 μm size ~ 10 μm - 440 μm
15 - 25 μm	6.1 %			
> 25 μm	2.5 %			
R_a = 0.276 μm, R_q = 0.548 μm, $\beta_{E,max}$ ≈ 14.6		R_a = 0.466 μm, R_q = 0.646 μm, $\beta_{E,max}$ ≈ 12.9	$\beta_{E,max}$ ≈ 4	R_a = 0.295 μm, R_q = 0.489 μm, $\beta_{E,max}$ < 4

It is remarkable that the highest $\beta_{E,max}$ values were obtained for particles (≤ 15) and scratch-like protrusions (≤ 13). Particles have been found on the surfaces more often than scratches. In some cases one could find up to 55 particles in an area of 500×500 μm² (Fig. 5.7(a)), but surely most of them could be removed by HPR and/or DIC. In contrast, scratches with high and sharp ridges formed by some material shift from electropolished surface must be avoided during the handling of cavities (Figs. 5.7(b) and 5.7(d)).

Other irregularities found on the flat Nb surfaces are hill-like structures (Fig. 5.7c) and round holes which are most probably caused by foreign material inclusions which might lead to reduced or enhanced electrochemical polishing. According to their comparably smooth edges and low $\beta_{E,max}$ values (≤ 4), holes and hill-like structures are not considered as sources of EFE as long as they do not have sharp edges.

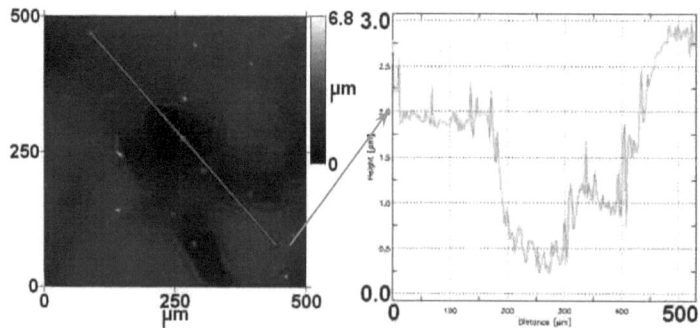

Fig. 5.8: Optical profile and line scan of a flat Nb sample with grain boundary steps.

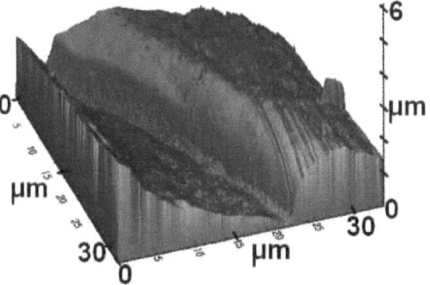

Fig. 5.9: AFM result showing a grain boundary step.

The grain structure of the flat Nb samples shown in the Fig. 5.8 was very similar to that of the curved half-cell ones. The average grain size of about 140 µm and a maximum step height up to 1.55 µm was determined from the optical profiles. AFM zoom scans into selected grain boundaries resulted in more precise values of the step edge radius (< 0.78 µm) as shown in the Fig. 5.9. Accordingly, the grain boundaries on the flat Nb samples also provided only low $\beta_{E,max}$ values ($\leq$ 4) which should not cause EFE in high gradient superconducting cavities. On the other hand, the surface defects with low electric field enhancement might reduce the magnetic field limits by magnetic field enhancement and cause high-field Q-drop and quenches [Kno99b].

5.3. Correlated FE/SEM/OP investigations of the flat EP Nb samples

In order to understand the nature of emitting defects and to study the influence of their geometry on EFE, a FESM-SEM correlating study of typically prepared flat Nb samples have been done. In order to have an overview of density distribution of emitter on the sample surface and their onset field, voltage scans at constant FE current of 1nA have been performed in the first instance (Fig. 5.10). Areas for the scans were chosen from the previous OP investigations (for example Fig. 5.7) or arbitrary. Voltage scans have been done at constant anode to cathode distance of about 50 µm with increasing of maximum field in steps starting from 80 MV/m. In some cases as soon as first emitters were found, further steps at higher field were not performed to avoid any modifications of the emitters.

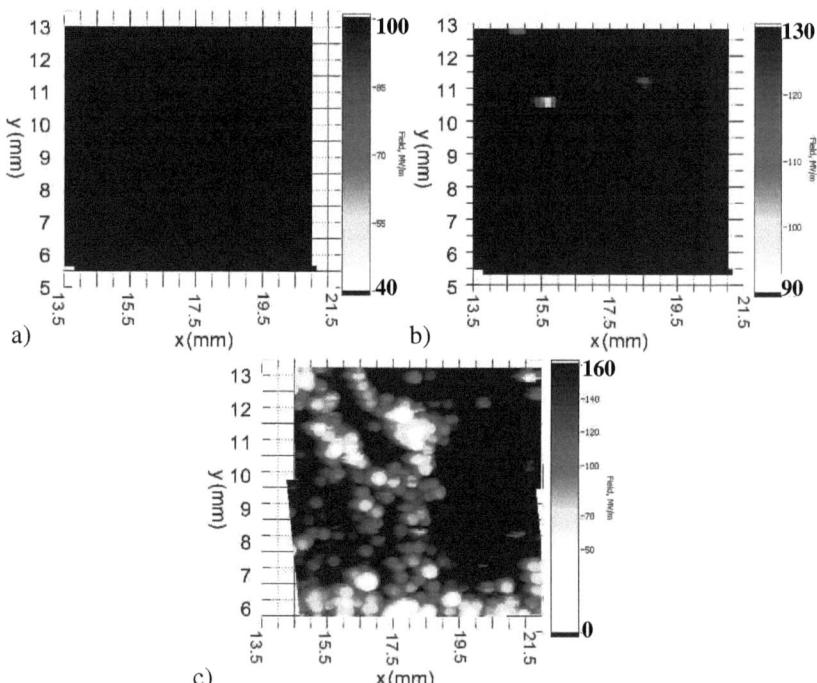

Fig. 5.10: Regulated onset-field maps $E_{on}(x,y)$ for constant current of 1nA and gap $\Delta z \approx 50$ µm of the same area at 100 (a), 130 (b) and 160 MV/m (c).

The resulting field maps showed drastic exponential increase of emitter site density starting at around 80 MV/m of macroscopic surface field (Fig. 5.11). Such increase of emitter site density is governed by presence of many defects (mainly scratches) on the Nb surface.

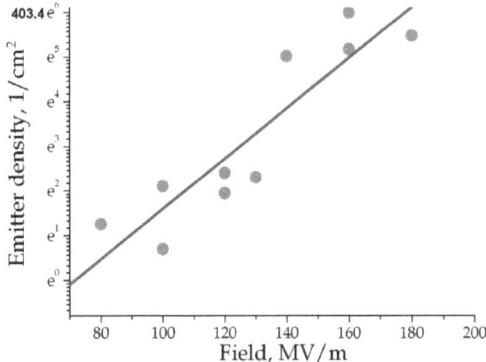

Fig. 5.11: Exponential increase of emitter site density with linear increase of electric field (activation field).

The second experimentally observed interesting effect is activation or switching on of emitters. It was discovered that many emitters start to be active at the field values much below the point where they have appeared for the first time. For example, the field map in the Fig. 5.10(c) taken at maximum of 160 MV/m contains many emitters with onset field lower that 60-70 MV/m, while scan at maximum 100 and 130 MV/m showed no and only 3 emitters on the surface correspondingly. The effect is found to be mostly irreversible. Once the emitter is switched on, it emits at lower field even during a next field-rise cycle after some time of passive non-emitting period. The activation field is typically 2 to 4 times higher that the following onset field of the emitters. Possible explanations for this phenomenon are an antenna effect, de- adsorption effect, or surface erosions due to microplasma. The basis of the antenna effect is most probably Nb surface oxide leading to creation of metal-insulator-metal (MIM) or metal-insulator-vacuum (MIV) structures of emitters [Ath84, Ath85, Xu95] (see Chapter 2). Due to the high local electric field (activation field), the oxide can be damaged by creation of conducting channels. The channels can have big aspect ration leading to increase of local field enhancement and finally to emission at much lower field (MIV case) or to be a connector between the bulk material and a particle sitting on the oxide leading to EFE from the particle (MIM case).

Real metal surfaces are covered by adsorbates from the air or by residual gases mostly. The transport of the samples was dust free, but the samples had long contact with water in the HPR machine and were exposed to the ambient air for many hours. In one hour, the cleanest metal surface even at 10^{-10} mbar can be covered by an atomic monolayer [Zei91] of adsorbates. The adsorbates on the metal surface lead to instability of FE due to their high mobility at the room temperature [Zei88, Swa67]. It was also observed that the electric field initiate significant step-like jumps of FE current which are believed to be due to formation of adsorbates-enhanced tunneling configuration between adsorbates and FE structures [Yeo04]. Many works show changes of work function of materials by adsorbates [Cla68, Bel66, Bag08] what can also lead to changes of FE behavior. However, changes of work function alone cannot fully explain the FE behavior and it is often misleading. For example, decrease of FE current at reduced work function of materials due to adsorbates [Ehr61] or wise versa. S factor often has also unrealistically high (some meter square) or small (much below atom size) value, while in the classical FN theory the S factor corresponds to an emission area. Therefore, a theory of resonant tunneling through localized energy levels of adsorbed atoms on the metal surface can describe the effect as it was introduced by Duke and Alferieff [Duk67] and later by Gadzuk [Gad70]. Water with its strong dipole moment is considered to be responsible for a crucial EFE [Hal83]. To avoid it, a baking of metallic surfaces in UHV conditions is required.

The third explanation of the switching effect is related to micro-discharge or explosive-emission-induced erosion of the metal surface (Fig. 5.12) [Wan97]. Often, the micro-discharge leaves molten craters (Fig. 5.13) on a bigger area with rather sharp features as the initial ones. The primary FE sites can be metal protrusions as well as foreign particulates on the surface. Obviously, the process is irreversible. It is also believed that often strong rf high power processing used for the cavities treatment in order to avoid FE and can lead to creation of the permanent emitter sites on the surface of the cavities.

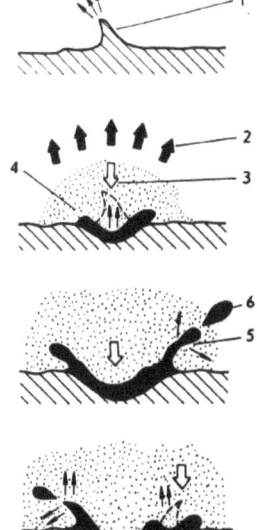

Fig. 5.12: Development of erosion on the metal surface in one cycle of explosive emission: 1 – primary FE site, 2 – electron current of explosive emission, 3 – ion pressure on the emission site, 4 – cathode plasma, 5 – growth of a new micro-point from the liquid metal, 6 – micro-drops remaining on the surface as metal bubbles [Wan97].

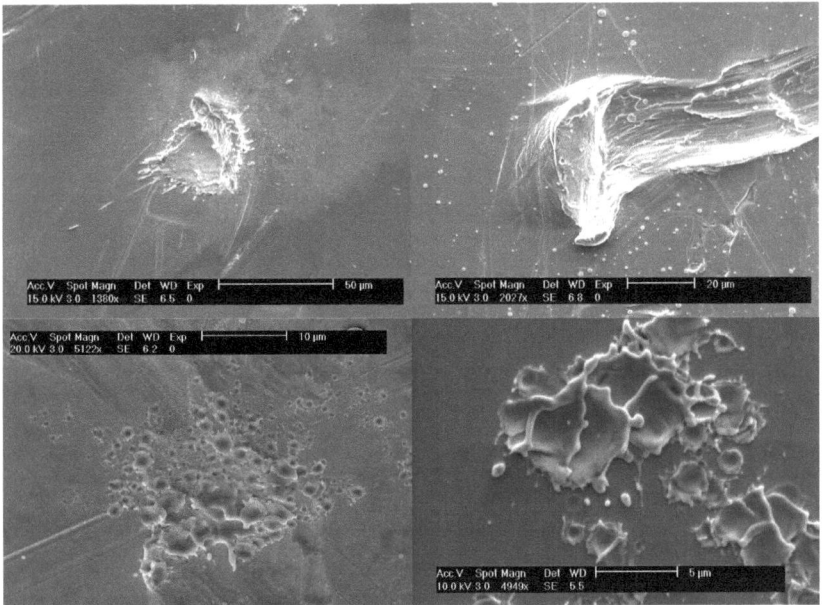

Fig. 5.13: SEM images of surface damages produced during the FE study by micro-discharges (erosions) observed on the flat EP Nb samples leading to permanent activation of emitters.

Considering the overall impact of the switching effect for the accelerator cavities, it is believed that heating [Mah94] and rf power [Wan97] might activate emitter too. It is despite that FE in the classical FN theory is independent of frequency. The basic phenomenon of electron FE as a source of dark current and as a trigger of rf breakdown in accelerator structures is well-established. One thing that is still missing is a deeper understanding of what exactly makes up the value of the enhancement factor β_E and how it is to be correlated with, and ultimately predicted from an observable surface condition. When many similar surface emitters are present, is the measured β_E the result of the superposition of these emitters or is it always dominated by one of them? One of the difficulties to answer this question is obvious overlapping of many possible sources of FE on real metal surfaces. It is always sum of geometrical defects, foreign inclusions, and surface conditions.

In order to have higher clarity concerning this, local FE investigations of the individual defects were performed. Onset field E_{on}, field enhancement factor $\beta_{E,FN}$ and S_{FN} factor were derived from corresponding FN plots assuming work function of 4.0 eV for Nb. Afterwards it was tried to correlate the FE properties of the defects to their geometry using results of OP and high resolution SEM investigations. In the actual setups, positioning accuracy between FESM and SEM (OP) was around 200µm. Geometrically calculated electric field enhancement $\beta_{E,G}$ was compared to $\beta_{E,FN}$ one derived from FN plots. An example of a correlated scratch-defect, which showed stable FN behavior after activation at 160 MV/m is shown in the Fig. 5.14. Geometrically calculated field enhancement factors of rims of the scratch are from $\beta_{E,G} \approx 50$ to 85. After the activation, the defect started to emit at much lower field of 35 MV/m. Field enhancement factor of the activated emitter was extracted from FN plot (Fig. 5.15) and is in a good correlation with the geometrical one: $\beta_{E,FN} \approx 60$. Emitting area of the defect is $S_{FN} \approx 70$ nm^2 what is smaller than it could be estimated from geometry of the defect. Discrepancy is most probably due to limited resolution of OP or interference of emission from many emitters. Comparison of the geometrical $\beta_{E,G}$ and field enhancement extracted from FN $\beta_{E,FN}$ shows good correlation in case of large clear defects. However, the activation or switching effect cannot be explained by geometrical considerations. Most probably, activation in this case is governed by surface oxide (MIM or MIV case) because no erosion (Fig. 5.14) and no foreign inclusions have been detected by EDX analysis (Fig. 5.15).

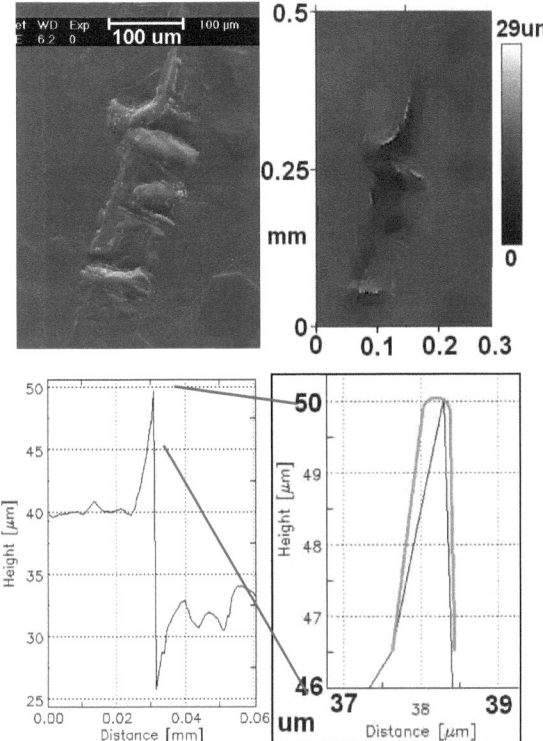

Fig. 5.14: SEM and OP images of a scratch with sharp rims smaller than 0.2 μm curvature radius (fitted by the green line) and up to 17 μm height.

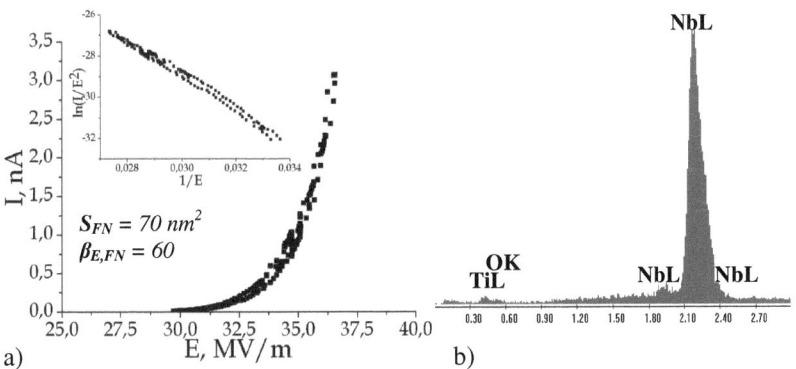

Fig. 5.15: (a) I-E and FN plot (inset) of the defect shown in figure 5.14 activated at 160 MV/m and showing stable FN emission afterwards starting at E_{on} = 35 MV/m. (b) Result of EDX analysis in the area of the defect showing no foreign inclusions.

In case of other small and/or denser distributed defects it was difficult to identify the emitter location due to their complex geometry (Fig. 5.16) and the mentioned limited positioning accuracy of the setups.

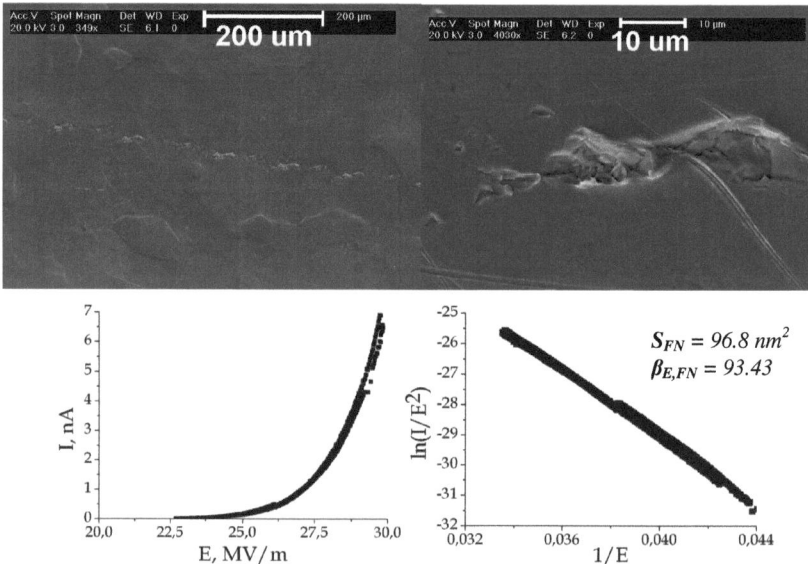

Fig. 5.16. SEM pictures (top) of a long scratch with complex shape of sharp rims activated at 160 MV/m and showing stable FN emission (bottom) afterwards starting at E_{on} = 27.2 MV/m.

In case of surface irregularities, it was interesting to find that the FN curves of most of the emitters were rather straight, i.e. showing stable FN behavior. It indicates good contact of emitters with the surface (Fig. 5.15 – 5.17) and insignificant influence of adsorbates. In other cases, FN curves showed unstable behavior indicating improper connection of the emitters or partial destruction of them by FE current with sequential increase of onset field (Fig. 5.18). The actual study was focused mainly on the surface defects despite some particulates that also were found on the surface. FE investigations of last ones showed rather unstable FN behavior. The EDX analysis in case of particulates showed presence of Al, Si, S, Cl, and Ca which are coming from the chemical polishing and the aluminum caps most probably. Some contamination during the FESM investigations cannot be excluded. For example, small bubbles of tungsten (Fig. 5.19) found close to some defects and indicate a partial destruction of the FESM needle-anodes due to sparks (highly likely during voltage scans due to limited speed of the voltage regulation system of the FESM).

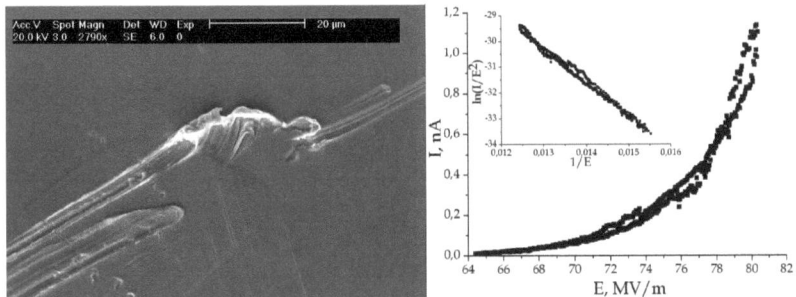

Fig. 5.17: SEM picture of a scratch activated at E_{act} = 120 MV/m and showing stable FN emission afterwards starting at E_{on} = 55 to 80 MV/m (S_{FN} = 1.8 nm^2, $\beta_{E,FN}$ = 42.3).

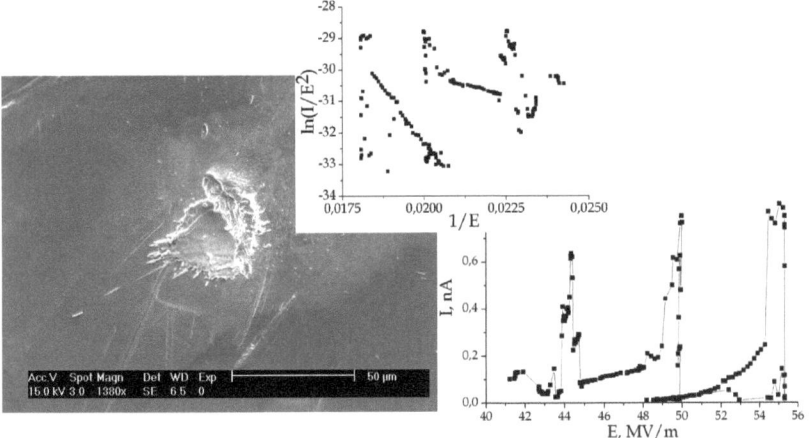

Fig. 5.18: SEM picture of a destroyed emitter activated at $E_{act} \approx$ 140 MV/m and showing stepwise destruction with increase of E_{on} from 43.9 to 54 MV/m (at 0.5 nA) with change of S_{FN} = 32300 nm^2, 0.12 pm^2 and 836 nm^2 and $\beta_{E,FN}$ = 43.45, 201.8 and 40.43 corr.

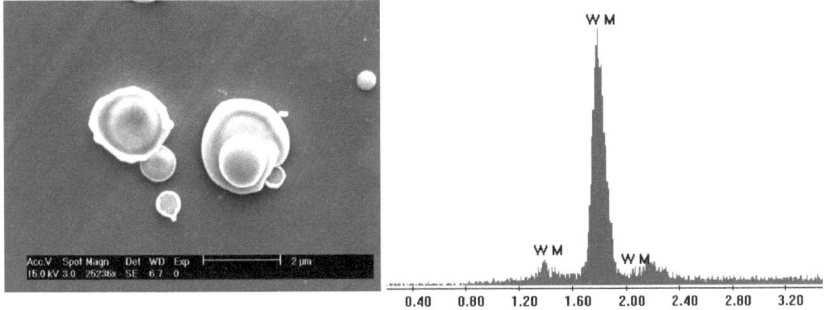

Fig. 5.19: Bubbles of W on the surface close to a defect and corresponding EDX spectrum.

Analyzing the overall impact of different geometrical defects of Nb surfaces on the performances of the Nb cavities one can compare the experimentally measured onset field (of already activated emitters) with the field enhancement $\beta_{E,FN}$ (from FN plots) of the emitters. Knowing design values of accelerating gradients of XFEL and ILC and considering an increase of macroscopic surface electric field due to ratio of peak to accelerating field E_p/E_{acc} one can find out tolerable values of field enhancement factor of defects for these machines. It is obvious that defects with $\beta_E \geq 50$ for XFEL and $\beta_E \geq 20$ for ILC should be completely avoided (see Fig. 5.20). That means the surface quality of the investigated samples is far not enough to fulfil requirements to the accelerator cavities even for the XFEL. It indicates the importance of a careful surface handling as a step to success in the accelerator physics.

It is difficult to judge about a tolerable size of defects on the surface as it was done previously [Dan09] because mainly sharpness of the defects (ratio of height to curvature radius) makes field enhancement and not their size alone. Defects with nanometer sharp rims would cause the highest EFE. Therefore, a different approach was made here by plotting onset field of the defects against their field enhancement (Fig. 5.20) and not the size. However, it makes more difficult the final optical inspection of the cavities, for example Aderhold [Ade09, Ade10] does it with respect to identification of probable places of EFE and quenches.

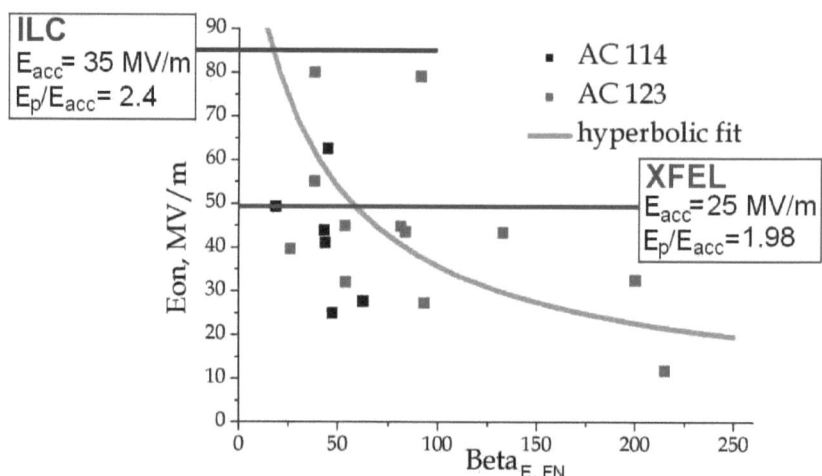

Fig. 5.20: Onset electric field (activated, for 1 nA FE currents) vs. field enhancement factor $\beta_{E,FN}$ (measured from FN plots) of all emitters found on AC114 and AC123 Nb samples with respect to the design values of E_{acc} and E_p fields of XFEL and ILC showing maximum tolerable β_E values.

In conclusion, a big effort was made to investigate the EFE from surface defects on EP/HPR Nb required for high gradient superconducting accelerator modules. It was attempted to obtain the correlation between geometry of defects and their FE properties. Optical profilometry was found to be a suitable tool for fast surface quality control of EP Nb surfaces with regards to the defects of some micrometer size. In contrast smaller defects require techniques with higher resolution such as AFM and SEM. Remaining particulates on the surface with $\beta_{E,MAX} > 15$ can be removed by HPR and DIC techniques, while geometrical defects have to be prevented by a more careful handling during production and assembly of the modules. It was found that the exponential increase of emitter site density on EP/HPR Nb surfaces and activation of emitter at 2-4 times higher fields as following onset field. Due to rather high activation field of emitter, it appears to be safe to operate the cavities up to around 80 MV/m, but activation by baking and influence of rf power should be taken into account and further investigated. Pits with crater-like centers and sharp rims have been found on the real cavity surfaces and hint for problems with chemical surface treatments. Grain boundaries with step height of some micrometer, hills, and holes do not cause EFE but have to be considered as sources of quenches and magnetic field limitations. Magnetic field enhancement on such defects has to be further investigated. It was shown that defects with $\beta_E \geq 50$ for XFEL and $\beta_E \geq 20$ for ILC should be completely avoided for successful suppression of FE load.

More detailed correlation studies on the samples with less defect density, better positioning accuracy, and in-situ SEM in FESM with at least one µm resolution are planned in near future.

Summary and outlook

The studies presented in this thesis are concerned with field emission (FE) investigations of the structured cathodes based on carbon nanotubes (CNT) and metallic nanowires (MNW) required for cold cathode applications, as well as investigation of parasitic FE from Nb surfaces and photocathodes required for particle acceleration machines.

The FE measurements were performed with the FE scanning microscope (FESM) for spatially resolved measurements and the integral measurement system with luminescent screen (IMLS) for applicational issues. Correlations between FE results and surface morphology/defects were made by means of optical profilometer (OP), atomic force microscope (AFM) and scanning electron microscope (SEM) including energy dispersive x-ray analysis (EDX). Several repairs/improvements of the FESM microscope were carried out including replacement of the high voltage insulator, replacement of the ball bearings and some parts of sliding stages, installation of the overload protection circuit for ammeter (for local I-V measurement), etc. In order to reach higher field levels of the FESM, required especially for parasitic FE investigations from Nb surfaces, the 5 kV power supply was replaced by more modern 10 kV one. The LabView programs of the FESM were correspondingly changed and some new useful functions, significantly helping measurements were added. Few more tasks concerning the FESM improvements, however, have to be solved in the future. It requires a protection of the ammeter input circuits during voltage scans and a reliable control of FUG power supply by a modern ammeter. The signal registration of the IMLS was significantly improved comparing to the previous version. Serial resistors were used for the cathode protection against discharges and fast current readout, as well as a resistive voltage divider for the high voltage registration. The IMLS was correspondingly equipped with a data acquisition system based on the self-developed software (using LabView) and the analog to digital converter (ADC) converter. It allows registration of full signals of the current and voltage in the dc and pulse modes as well as their time dependence during long-term tests. The triode configuration of the system, however, has to be still improved. For registration of luminescent screen images of the IMLS, it is suggested to replace the actual camera by a newer one with a better focusing and a LabView compatible interface. There is an idea to incorporate the acquisition of images into the LabView program and store the images as time- and I-V-referenced data for a faster processing and a clear correlation between them.

A new system, the novel ultra high vacuum (UHV) scanning anode FE microscope (SAFEM) has been designed and built within this thesis work. It is presented rather more

Chapter 6 Summary and Outlook 99

detailed as it was a significant part of the work. The SAFEM has been developed in order to ensure low FE photocathodes required for electron guns of free-electron lasers like FLASH or the future European XFEL, and to investigate dark current sources in details, and to improve cathode-handling techniques. Imperfect photocathode regions with enhanced field emission (EFE) and their contact area to the rf cavity are considered as main dark current sources at typical macroscopic surface electric fields of about 40-60 MV/m. The microscope is designed to achieve dc surface fields of at least 200 MV/m and provides the localization of the field emitters with a spatial resolution of about 1 µm in UHV of at least 10^{-10} mbar. The design (performed using SolidWorks 3D CAD software package), control software (self-developed using LabView), and initial assembly of the SAFEM are completed. First vacuum test of the SAFEM, however, suffered from the strong gassing of the Micos sliding stages, hindering achievement of the specified value of the vacuum level. The sources of the gassing had been established, following by replacement of some parts. After successful vacuum tests and first commissioning of the SAFEM, it will be finally included into the preparation and systematic quality control set-up of the photocathodes at DESY.

Varying arrays of CNT columns and blocks for cold cathode application were fabricated by two different chemical vapor deposition (CVD) methods. Applying the atmospheric pressure CVD method developed at BSUIR, the columns of CNT were preferentially grown on flat Si substrates using a ferrocene/xylene mixture as volatile catalyst source. Growth times of 30 s (2 min) was resulted in vertically aligned columns of about 20 µm (50 µm) net heights, consisting of densely packed entangled CNT. The columns are embedded in a cloud-like floor of shorter CNT still grown on the SiO_2. The structured arrays of the CNT columns form 4 quadrants of 2x2 mm^2 size with round patches and pitch to diameter ratio of 160/50, 100/30, 100/50 and 100/10 µm, respectively. The water assisted CVD method of the CNT deposition realized in TUD showed 100% selectivity of growth comparing to the atmospheric pressure CVD method. Structured arrays of aligned CNT blocks were fabricated on a flat p-type Si substrates by applying a bimetallic (aluminum and patterned iron) catalyst and the water assisted CVD. For growth times of 5 to 10 min, the resulting multiwall CNT formed vertically aligned arrays of uniform rectangular blocks of 50, 100 or 140 µm width, 150 to 600 µm height, and 230, 250 and 300 µm pitch. Some of the CNT block arrays were CVD-coated with a thin TiO_2 layer (< 50 nm) to study current stabilization effects and influence of it on the onset field and the maximum current capability of the blocks. It is remarkable that the TiO_2-coated blocks are sharpened and partially tilted due to their high aspect ratio. Well-aligned FE from nearly 100% of the blocks and columns at electric field <10 V/μm was observed. High current capabilities of the columns up to mA and stable currents up to 300 (100) μA for pure (TiO$_2$ coated) CNT blocks, were achieved. Such high current capability of long CNT column is correlated to improved electrical contact and mechanical stability due to the cloud-like floor of CNT on the substrate. It reveals an importance and need of an improved contact interfaces for both types of the CNT structures. Current capabilities from individual patches of both types of materials as well as the derived field enhancement factors confirm the presence of multiple CNT emitters per patch. The recovering effect due to the redundancy and the multiple emitters confirmed our strategy to improve the FE performance of the cathodes. There is some evidence that TiO$_2$ coating stabilize FE at low currents (I < 10μA per patch), showing less stepwise deactivation of emission and more FN like behavior of emission. It can be explained by smoothing of the blocks and binding of individual (probably even loosen) CNT on the blocks by TiO$_2$ layer. However, it was found out also that TiO$_2$ coating lead to factor of 2-3 reduction of maximum currents limits of individual blocks comparing to pure ones as well as factor of two higher sensitivity of TiO$_2$-coated CNT to oxygen. Integral FE measurements with luminescence screen (IMLS) and processing under N$_2$ and O$_2$ exposures of up to 3×10^{-5} mbar demonstrated fairly homogeneous current distribution and long-term stability of the CNT cathodes. However, the uniformity over a full cathode has to be still improved by adjustment of geometry and better uniformity of individual patches for both types of the cathodes. Use of the actual CNT cathodes for such application, which require high quality stable and homogeneous cathodes, is not possible, yet. However, some application like simple x-ray sources can be already built based on this material. It was shown that values of field enhancement factor β_{eff} could not be explained by the ratio of the patch height and the CNT radius but hint for varying mutual shielding effects within the columns and field enhancement by outstanding individual CNT. FE is determined by structure of the upper edge of the patch rather than by geometry of the full patch. It makes aspect ratio and correspondingly full height of the structures nearly irrelevant for the field enhancement, indicating no need in such high structures. For future improvements of the cathodes, the height should be reduced to about 20 μm to avoid their tilting and difference in cross-section and height. The pitch of the arrays should be at least four times of the block width and three times of their height to improve alignment and decrease their interference and mutual shielding. A more detailed FE study of such CNT cathodes with improved geometry and uniformity of the blocks as well as FE spectroscopy measurements for better understanding of coating effects are to be done in the future.

Cathodes containing regular patch arrays of random metallic nanostructures, as an interesting alternative to the CNT, were fabricated by an electrochemical deposition of Au and Pt nanowires (NW) into ion track-etched templates and were systematically investigated with the SEM and FESM. FE with about 90% efficiency was achieved. The best result in terms of efficiency and alignment of the FE sites was achieved for double-spaced patches of 50 µm diameter with Au-NW at number densities of 10^7 cm^{-2}. Multiple nanostructures per patch contribute to high emission efficiency of the cathodes. High field enhancement and low onset fields are fairly well correlated for solitary NW, but the achievable maximum currents carrying capability of individual patches, however, strongly varied between 40 nA and 90 µA for both types of nanostructures. First correlation studies between FESM maps and SEM images of selected patches revealed the partial melting and successive destruction of solitary and cluster NW as a main cause for the obtained current limits. Thermal model calculations, SEM-FESM-SEM correlation studies and comparison of the results of Au and Pt NW patches revealed geometrical constrictions in the NW contact region as a main source of the observed limitations. These results have given considerable hints for a modification of the fabrication steps to improve FE homogeneity, current strength and stability of the structured metallic NW cathodes.

The EFE from particulate contaminations or surface irregularities is one of the main field limitations of high gradient superconducting Nb cavities required for future accelerators the XFEL and the ILC. Therefore, systematic measurements of the average and local surface roughness of typically prepared Nb samples, some of which were cut out of a real nine-cell polycrystalline Nb cavity have been carried out. Altogether six flat and five curved Nb samples were tested. The flat ones with Ø 26.5 mm were machined from the standard for the SRF cavity production high-purity Nb sheets (RRR > 300) and welded to support rods. The curved samples were cut out of the nine-cell cavity, which suffered from the strong quenching at moderate accelerating field of 16 MV/m. These samples were taken from different interesting surface areas: two (one) from the flat part of the lower (upper) half-cell and one each from the iris and equator weld regions. All samples have obtained comparable EP and BCP etching as well as HPR. Artificial scratches of the flat samples, as a result of mechanically unsafe storage conditions, were used for the correlation study between geometry of the defects and their FE properties. By means of the OP combined with the AFM, large scanning ranges as well as high-resolution zooms into found defect areas were obtained. The OP was found to be a suitable tool for fast surface quality control of EP Nb surfaces with respect to defects of some micrometer size. In contrast, smaller defects require techniques with higher resolution such as AFM and SEM. After HPR of selected flat Nb samples at DESY, correlated FESM and high-resolution SEM investigations have been done in order to reveal the influence of the defect geometry on the expected EFE and accordingly on the performances of the accelerator modules. Particulates and scratches were identified as potentially stronger field emitters than grain boundaries, round hills and holes. Remaining particulates on the surface with $\beta_{E,MAX} > 15$ can be removed by HPR and DIC techniques, while geometrical defects have to be prevented by a more careful handling during production and assembly of the modules. The exponential increase of the emitter site density on EP/HPR Nb surfaces and activation of emitter at 2-4 times higher fields as following onset field were found. Possible explanations for this phenomenon are the antenna effect, de- adsorption effect, or local surface erosions due to strong breakdowns. Due to rather high activation field of emitters, it might be safe to operate the cavities up to around 80 MV/m, but activation by baking and influence of rf power should be taken into account and further investigated. The pits with crater-like centers and sharp rims have been found on real cavity surfaces and it gives us a hint for problems with chemical surface treatments. Grain boundaries with step height of some micrometer, hills, and holes do not cause EFE but have to be considered as sources of quenches and magnetic field limitations. Magnetic field enhancement on such defects, therefore, has to be further investigated. It was shown that defects with $\beta_E \geq 50$ for XFEL and $\beta_E \geq 20$ for ILC should be completely avoided for successful suppression of parasitic FE load of the accelerator modules. More detailed correlation studies on samples with less defect density, better positioning accuracy, in-situ SEM in FESM with at least 1 µm resolution, and investigations of baking effect are to be done in the future.

Acknowledgements

This has been a long way, and I have many people to thank for their help, support, and encouragement.

I would first like to thank my supervisor, Prof. Dr. Günter Müller for his inspiring guidance, constant availability, encouragement, patience, and professional as well as personal support. He continually stimulated my scientific growth and greatly assisted me with scientific writing. I am grateful to him for his invitation to work and study in Germany.

I would like to thank Jahan Pouryamout for constant availability and readiness to help me in minor and major problems and for his valuable hints especially concerning vacuum techniques.

I thank my colleagues from the laboratory over the years: Vitali Sakharuk, Dmitry Solovei, Benjamin Bornmann, Stefan Lagotzky, Felix Jordan, and Pavel Serbun for great help with measurements and for constructive discussions.

It is impossible imagine this work without cooperation with counterparts. I would like to thank Prof. Dr. Vladimir A. Labunov, Alena L. Prudnikava, Jury P. Shaman, Prof. Dr. Gennadii Gorokh, Prof. Dr. Alexander G. Smirnov at BSUIR, Minsk for providing CNT samples and for given help to come to German University. I thank Prof. Dr. Jörg. J. Schneider, Dr. Jörg Engstler, and Dr. Ravi K. Joshi at TUD, Darmstadt for providing CNT samples. I thank my colleagues from DESY, Mr. Axel Matheisen for providing flat Nb samples, Dr. Detlef Reschke for help with high pressure rinsing, Dr. Xenia Singer for providing curved Nb samples, Dr. Klaus Flöttmann and Dr. Sven Lederer for help and cooperation regarding the construction and building of the SAFEM. I am grateful to Dr. Christina Trautmann, Dr. Maria Eugenia Toimil Molares, Dr. Thomas Walter Cornelius, Dr. Shafqat Karim, Ina Alber, Sven Müller, and Markus Rauber at GSI, Darmstadt for providing MNW samples. And I am grateful of course to all the colleagues for a number of fruitful discussions.

I thank Dr. R. Heiderhoff at the Electrical Engineering Department of the University of Wuppertal for support with high resolution SEM and EDX analysis.

Special thank to Prof. Dr. Ronald Frahm for his kind agreement to review this work.

It is my greatest pleasure to thank my parents Navitski Mikalai and Navitskaya Raisa for their invaluable efforts in stressing their attention to my education. I thank my parents-in-law Rakovich Mikhail and Tatsiana for their unrestricted support and encouragement. I would like to thank my wife Iryna and my daughter Alina for patience, love, and great support. They always motivate me to work harder and do my best. I hope that now I can spend more time with them.

I would like to thank my friends Alida Hübner, Oleg Sergeev, and Lizaveta Sergeeva for their great support during my stay in Germany.

Financial support for the work from the Helmholtz-centers DESY and GSI, the Helmholtz Alliance "Physics at the Terascale," and the German Federal Ministry of Education and Research (BMBF) is highly appreciated.

References

[Ade09] S. Aderhold, Proceedings of the 14th International Workshop on RF Superconductivity, Berlin, Germany, TUPPO035, p. 238 (2009).

[Ade10] S. Aderhold, Proceedings of IPAC'10, Kyoto, Japan, WEPEC005, p. 2896 (2010).

[Ado08] J. Ado, G, Dresselhaus, M. S. Dresselhaus, „Carbon nanotube. *Advanced Topics in the Synthesis, Structure, Properties and Applications*," Springer, Berlin, ISBN: 978-3-540-72864-1 (2008).

[Alp64] D. Alpert, D.A. Lee, E.M. Lyman, and H.E. Tomaschke, J. Vac. Sci. Technol. **1**, 35 (1964).

[Ana] Olympus Soft Imaging Solutions GmbH, www.olympus-sis.com

[Ant99] C.Z. Antoine, Proceedings of the 9th International Workshop on RF Superconductivity, Santa Fe, New Mexico, USA, TUA008, 109 (1999).

[Apk52] U. Apker and E. A. Taft, Phys. Rev. **88**, 1037-1038 (1952).

[Ast07] ASTM D5127-99 and update D5127-07: "*Standard Guide for Ultra pure Water in the Electronics and Semiconductor Industry*" (1999 + 2007).

[Ath84] C. S. Athwal and R V Latham, J. Phys. D: Appl. Phys. **17**, 1029 (1984).

[Ath85] C. S. Athwal, K. H. Bayliss, R. Calder, and R. V. Latham, IEEE Trans. Plasma Sci. **13**, 226 (1985).

[Bac00] A. Bachtold, M. S. Fuhrer, S. Plyasunov, M. Forero, E. H. Anderson, A. Zettl, and P. L. McEuen, Phys. Rev. Lett. **84**, 6082 (2000).

[Bag08] P. S. Bagus, D. Käfer, G. Witte and C. Wöll, Phys. Rev. Lett. **100**, 126101 (2008).

[Bel66] A. Bell and R. Gomer, J. Chem. Phys. **44**, 1065 (1966).

[Ben67] C. J. Benette, L. W. Swanson, and F.M. Charbonnier, J. Appl. Phys. **38**, 634 (1967).

[Ber00] S. Berber, Y.-K. Kwon, and D. Tomanek, Phys. Rev. Lett. **84**, 4613 (2000).

[Bon01] J. M. Bonard, N. Weiss, H. Kind, T. Stockli, L. Forro, K. Kern, and A. Chatelain, Adv. Mater. **13**, 184 (2001).

[Bon02] J.-M. Bonard, A. D. Kenneth, F. Bernard, and C. Klinke, Phys. Rev. Lett. **89**, 197602 (2002).

[Bon03] J.-M. Bonard, C. Klinke, A.D. Kenneth, and F. C. Coll, Physical Review B **67**, 115406 (2003).

[Bor07] V. S. Bormashov, A. S. Baturin, K. N. Nikolskiy, R. G. Tschesov, and E. P. Sheshin, Surf. Interface Anal. **39**, 155 (2007).

[Bor10] B. Bornmann, S. Mingels, A. Navitski, D. Lutzenkirchen-Hecht, and G. Muller, Proc. of the 23^d Int. Vacuum Nanoelectronics Conf. IVNC2010, Palo Alto, USA, 4-1, p. 20 (2010).

[Bow02] C. Bower, W. Zhu, D. Shalom, D. Lopez, L. H. Chen, P. L. Gammel, and S. Jin, Appl. Phys. Lett. **80**, 3820 (2002).

[Bro10] J. Browning, L. Matthews, J. Watrius, M. Eaton, and N. Kumar, Proc. of the 23^d Int. Vacuum Nanoelectronics Conf. IVNC2010, Palo Alto, USA, 3.1, p.11 (2010).

[Bsu] Belarusian State University of Informatics and Radioelectronics, BSUIR, www.bsuir.by.

[Bur53] R. E. Burgess, H. Kroemer, and J. M. Houston, Phys. Rev. **90**, 515 (1953).

[Bus10] K. H. J. Buschow, R. Cahn, M. Flemings, B. Ilschner, E. Kramer, S. Mahajan, and P. Veyssiere, *"Encyclopedia of Materials: Science and Technology,"* ISBN: 978-0-08-043152-9, Pergamon, last update January 2010.

[Che04] C.-W. Chena, M.-H. Leeb, and S.J. Clark, Applied Surface Science **228**,143 (2004).

[Chr88] J. R. Christman, *"Fundamentals of solid state physics,"* John Wiley & Sons, New York, ISBN: 0-471-81095-9 (1988).

[Cio06] G. Ciovati, Proc. of LINAC 2006, Knoxville, Tennessee USA, FR1004, 818 (2006).

[Cit79] A. Citron, G. Dammertz, M. Grundner, L. Husson, R. Lehm, and H. Lengeler. Nucl. Instr. and Meth. **164**, 31 (1979).

[Cla68] H. E. Clark and R. D. Young, Surf. Sci. **12**, 385 (1968).

[Com] COMSOL Multiphysics® 3.5a, www.comsol.com.

[Cui01] J. B. Cui, J. Robertson, and W. I. Milne, J. Appl. Phys. **89** 5707 (2001).

[Dan06] A. Dangwal, D. Reschke, and G. Müller, Physica C **441**, 83 (2006).

[Dan07a] A. Dangwal, G. Müller, D. Reschke, K. Floettmann, and X. Singer, J. Appl. Phys. **102**, 044903 (2007).

[Dan07b] A. Dangwal, G. Müller, F. Maurer, J. Brötz, and H. Fuess, J. Vac. Sci. Technol. B **25**, 586 (2007).

[Dan08] A. Dangwal, C. S. Pandey, G. Müller, S. Karim, T. W. Cornelius, and C. Trautmann, Appl. Phys. Lett. **92**, 063115 (2008).

[Dan09] A. Dangwal, G. Müller, D. Reschke, and X. Singer, Phys. Rev. ST Accel. Beams **12**, 023501 (2009).

[Dav68] D. K. Davies, M. A. Biondi, J. Appl. Phys. **39**, 2979 (1968).

[Dea00] K. A. Dean and B. R. Chalamala, Appl. Phys. Lett. **76**, 375 (2000).

[Dea01] K. A. Dean, T. P. Burgin, and B. R. Chalamala, Appl. Phys. Lett. **79**, 1873 (2001).

References

[Des] Das Deutsche Elektronen-Synchrotron DESY, http://desy.de.

[Die71] H. Diepers, O. Schmidt, H. Martens and F. S. Sun., Phys. Lett. **37A**, 139 (1971).

[Din06] F. Ding, K. Bolton, and A. Rosen, J. Electronic Materials **35**, 207 (2006).

[Dio08] M. Dionne, S. Coulombe, J.-L. Meunier, IEEE transactions on electron devices **55**, 1298 (2008).

[Duk09] P. J. Duke, "Synchrotron radiation," Oxford University Press, ISBN 978-0-19-851758-0 (2009).

[Duk67] C.B. Duke, and M.E. Alferieff, J. Chem. Phys. **46**, 923 (1967).

[Ehr61] G. Ehrlich, F. G. Hudda, J. Chem. Phys. **35**, 1421 (1961).

[Ele] EleNet 6.18.1, http://www.infolytica.com.

[Fan99] S. S. Fan, M. G. Chapline, N. R. Franklin, T. W. Tombler, A. M. Cassell, and H. Dai, Science **283**, 512 (1999).

[Fis66] R. Fischer and H. Neumann, Fortschr. Phys. **14**, 603 (1966).

[Fla] Free-electron laser FLASH, http://flash.desy.de.

[Flö] K. Flöttmann,"*ASTRA User Manual*", www.desy.de/~mpyflo/Astra documentation.

[For99] R. G. Forbes, J. Vac. Sci. Technol. B **17**, 526 (1999).

[Fow28] R. H. Fowler and L. W. Nordheim, Proc. Royal Soc., London, **A119**, 173 (1928).

[Fra98] S. Frank, P. Poncharal, Z. L. Wang, and W. A. de Heer, Science **280**, 1744 (1998).

[Frt] FRT GmbH, http://www.frt-gmbh.com.

[Fug] FUG ElektronikGmbH, http://www.fug-elektronik.de.

[Fur05] G. Fursey, "*Field emission in vacuum microelectronics,*" Kluwer Academic, Plenum Publishers, New York, ISBN: 0-306-47450-6 (2005).

[Fur98] G. N. Fursey and D. V. Glazanov, J. Vac. Sci. Technol. B **16**, 910 (1998).

[Gad70] J.W. Gadzuk, Phys. Rev. B **1**, 2110 (1970).

[Gad93] J. W. Gadzuk, Phys. Rev. B **47**, 12832 (1993).

[Gai02] L. Gail, H.-P. Hortig (editors), "*Reinraumtechnik,*" Springer Verlag, ISBN 3-540-66885-3 (2002).

[Gar98] N. Garcia, M. I. Marques, A. Asenjo, and A. Correia, J. Vac. Sci. Technol. B **16**, 654 (1998).

[Giv95] E. I. Givargizov, V. V. Zhirnov, A. N. Stepanova, E. V. Rakova, A. N. Kiselev, and P. S. Plekhanov, Appl. Surf. Sci. **87**, 24 (1995).

[Gme70] L. Gmelin, "*Handbuch der anorganischen Chemie,*" Vol. 49 (Nb), Springer, Berlin (1970).

[Gom93] R. Gomer, "*Field emission and field ionization*," New York, ISBN: 1-56396-124-5 (1993).

[Goo59] R. H. Good and E. W. Müller, "*Field emission*," Vol. 21 of "*Handbuch der Physik*", edited by S. Flügge, Springer, Berlin, p.176-231 (1959)..

[Gos26] B. S. Gossling, Phil. Mag. **1**, 609 (1926).

[Göh00] A. Göhl, Dissertation WUB-DIS 2000-7, University of Wuppertal (2000).

[Grö00] O. Gröning, O. M. Küttel. C. Emmengger, P. Gröning, and L. Schlapbach, J. Vac. Sci. Techn. B **18**, 665 (2000).

[Grö01] O. Gröning, L.-O. Nilsson, P. Gröning, and L. Schlapbach, Solid State Electronics **45**, 929 (2001).

[Hab98] T. Habermann, Dissertation WUB-DIS 98-18, University of Wuppertal (1998).

[Hal83] J. Halbritter, IEEE Transactions on Electrical Insulation **EI-18**, 253 (1983).

[Har09] H. L. Hartnagel, O. Yilmazoglu, V. Litovchenko, A. Evtukh, and D. Pavlidis, Techn. Digest of the 22nd Int. Vacuum Nanoelectronics Conf. IVNC2009, Hamamatsu, IEEE Cat. No. CFP09VAC-PRT, p. 77-78 (2009).

[Hee95] W. A. de Heer, A. Chatelin, and D. Ugarte, Science **270**, 1179 (1995).

[Hig01] T. Higuchi, K. Saito, Y. Yamazaki, T. Ikeda and S. Ohgushiet, Proc. of the 10th Workshop on RF Superconductivity, Tsukuba, Japan, p. 431 (2001).

[Hsu05] D. S. Y. Hsu and J. L. Shaw, J. Appl. Phys. **98**, 014314 (2005).

[Ilc] International linear collider, http://www.linearcollider.org.

[Jan05] H. Jang-Hui, Dissertation, University of Hamburg (2005).

[Jan08] H. Jang-Hui, K. Flöttmann, and W. Hartung, Phys. Rev. ST Accel. Beams **11**, 013501 (2008).

[Jar10] J. D. Jarvis, N. Ghosh, B. L. Ivanov, J. L. Kohler, J. L. Davidson, and C. A. Brau, Proceedings of the Intern. Free Electron Conf. Malmö, Sweden, WEPB46 (2010).

[Jen09] K. L. Jensen, J. Appl. Phys., **107**, 014905 (2009).

[Jor08] A. Jorio, M. S. Dresselhaus, G. Dresselhaus, "*Carbon nanotubes: Advanced topics in the synthesis, properties and applications*," Topics of Appl. Phys. **111**, Springer. Berlin, Heidelberg, ISBN: 978-3-540-72864-1 (2008).

[Jos10a] R. K. Joshi, Ph.D. Thesis, Technical University of Darmstadt (2010).

[Jos10b] R. K. Joshi, J. Engstler, J. J. Schneider, A. Navitski, V. Sakharuk, and G. Müller, "CVD deposition of TiO_2 on CNT blocks," Mat. Sci. Journal (to be published) (2010).

[Jos10c] R. K. Joshi, O. Yilmazoglu, J.J. Schneider, and D. Pavlidis, J. Mater. Chem. **20**, 1717 (2010).

[Jun98] J. Jung, B. Lee, and J. D. Lee, J. Vac. Sci. Technol. B **16**, 920 (1998).

[Kar06] S. Karim S, M. E. Toimil-Molares, A. G. Balogh, W. Ensinger, T. W. Cornelius, E. U. Khan, and R. Neumann, Nanotechnology **17**, 5954 (2006).

[Kar07] S. Karim, Dissertation, Chem. Dep., University of Marburg, Germany (2007).

[Kei] Keithley Instruments GmbH, http://www.keithley.de.

[Kim06a] D. Kim, J. E. Bouree, S. Y. Kim, Appl. Phys. A **83**, 111 (2006).

[Kim06b] H. S. Kim, Y. C. Kim, D.-W. Kim, S. J. Ahn, Y. Jang, H. W. Kim, D. J. Seong, and K.W. Park, Microelectronic Engineering **83**, 962 (2006).

[Kne05a] P. Kneisel, G. R. Myneni, G. Ciovati, J. Sekutowicz, and T. Carneiro, Proc. of the 2005 Particle Accelerator Conference, Knoxville, Tennessee, p. 3991 (2005).

[Kne05b] P. Kneisel, G. Ciovati, G. R. Myneni, W. Singer, X. Singer, D. Proch, and T. Carneiro, Proc. of the 2005 Particle Accelerator Conference, Knoxville, Tennessee, p. 3955 (2005).

[Kne80] P. Kneisel, Proc. of the Workshop on RF Superconductivity, Karlsruhe, Germany. SRF80-2, p.27 (1980).

[Kne90] P. Kneisel, "Some History of Electropolishing of Niobium 1970 – 1990," TTC meeting, Frascati, Italy (2005).

[Kne95] P. Kneisel and B. Lewis, Proc. of 7th Workshop on RF Superconductivity, Gif sur Yvette, France, p. 311 (1995).

[Kne96] P. Kneisel and B. Lewis, Particle Accelerators **53**, 97 (1996).

[Kne97] P. Kneisel, K. Saito, and R. Parodi, Proc. of the 1997 Workshop on RF Superconductivity, Abano Terme, p. 463. (1997).

[Kno99a] J. Knobloch, IEEE transactions on applied superconductivity **9**, 1016 (1999).

[Kno99b] J. Knobloch, R.L. Geng, M. Liepe, and H. Padamsee, Proc. of 9th Workshop on RF Superconductivity, Santa Fe TUA004, p.77 (1999).

[Koe06] F.A.M. Koeck, and R.J. Nemanicha, Diamond and Related Materials **15**, 2006 (2006).

[Koj10] A. Kojima, T. Ohta, H. Ohyi, N. Koshida, Proc. of the 23^d Int. Vacuum Nanoelectronics Conf. IVNC2010, Palo Alto, USA, p.111, (2010).

[Kum10] N. Kumar, I. Yee, B. Fowler, R. Hellmer, B. Chalamala, Proc. of the 23^d Int. Vacuum Nanoelectronics Conf. IVNC2010, Palo Alto, USA, p.16, (2010).

[Kwo07] S. J. Kwon, Jpn. J. Appl. Phys. **46**, 5988 (2007).

[Lab] National Instruments Corp., http://www.ni.com

[Lab07] V. Labunov, B. Shulitski, A. Prudnikava, J. Shaman, A. Smirnov, A. Navitski, G. Mueller, Proc. of the 14th Int. Display Workshop, Sapporo, Japan, p. 2181 (2007).

[Lab09a] V. A. Labunov, B. G. Shulitski, A. L. Prudnikava, Y. P. Shaman, and A. S. Basaev, J. Soc. Information Displays **17**, 489 (2009).

[Lab09b] V. Labunov, A. Prudnikava, G. Gorokh, B. Shulitski, V. Sakharuk, A. Navitski, G. Müller, A. Basaev, Proc. of the 29th Int. Display Research Conf. Eurodisplay 2009, Rome, Italy, p. 248-250 (2009).

[Lat95] R. V. Latham, "High Vacuum Insulation: Basic Concepts and Technological Practice," Academic Press: London, ISBN: 0-12-437175-2 (1995).

[Lee05] J. H. Lee, S. H. Lee, W. S. Kim, H. J. Lee, J. N. Heo, T. W. Jeong, C. H. Choi, J. M. Kim, J. H. Park, J. S. Ha, H. J. Lee, J. W. Moon, M. A. Yoo, J. W. Nam, S. H. Cho, T. I. Yoon, B. S. Kim, and D. H. Choe, J. Vac. Sci. Technol. B **23**, 718 (2005).

[Lee10] H. Lee, J. Choi, C. Lee, J. Goak, N. Lee, Proc. of the 23^{d} Int. Vacuum Nanoelectronics Conf. IVNC2010, Palo Alto, USA, p.18, (2010).

[Lil04a] L. Lilje, C. Antoine, C. Benvenuti, D. Bloess, J. -P. Charrier, E. Chiaveri, L. Ferreira, R. Losito, A. Matheisen, H. Preis, D. Proch, D. Reschke, H. Safa, P. Schmüser, D. Trines, B. Visentin and H. Wenningeret, Nuclear Instruments and Methods in Physics Research A **516**, 213 (2004).

[Lil04b] L. Lilje, E. Kako, D. Kostin, A. Matheisen, W.-D. Möller, D. Proch, D. Reschke, K. Saito, P. Schmüser, S. Simrock, T. Suzuki, and K. Twarowski, Nuclear Instruments and Methods in Physics Research A **524**, 1 (2004).

[Lin09] M.-C. Lin, Techn. Digest of the 22nd Int. Vacuum Nanoelectronics Conf., Hamamatsu, IEEE Cat. No. CFP09VAC-PRT, p. 79-80 (2009).

[Lip04] D. Lipka, Ph.D. Thesis, Humboldt University Berlin (2004).

[Liu06a] J. Liu, J. L. Duan, M. E. Toimil-Molares, S. Karim, T. W. Cornelius, D. Dobrev, H. J. Yao, Y. M. Sun, M. D. Hou, D. Mo, Z. G. Wang, and R. Neumann, Nanotechnology **17**, 1922 (2006).

[Liu06b] K. X. Liu, C.-J. Chiang, and J. P. Heritage, J. Appl. Phys. **99**, 034502 (2006).

[Lu06] X. Lu, Q. Yang, C. Xiao, and A. Hirose, J. Phys. D: Appl. Phys. **39**, 3375 (2006).

[Lub] Lubricant Consult GmbH www.lubcon.com.

[Lys05] D. Lysenkov and G. Müller, Int. J. Nanotechnol. **2**, 239 (2005).

[Lys06] D. Lysenkov, Ph.D. Thesis, University of Wuppertal, WUB-DIS 2006-2 (2006).

[Lys07] D. Lysenkov, J. Engstler, A. Dangwal, A. Popp, G. Müller, J. J. Schneider, V. M. Janardhanan, O. Deutschmann, P. Strauch, V. Ebert, and J. Wolfrum, Small **3**, 974 (2007).

[Mah93] E. Mahner. Proceedings of the 6th Workshop on RF Superconductivity, CEBAF, Newport News, Virginia, USA p. 252 (1993).

[Mah94] E. Mahner, Part. Acc. **46**, 67 (1994).

[Mah95] E. Mahner, Ph.D. Thesis, University of Wuppertal, WUB-DIS 95-7 (1995).

[Man05] H. M. Manohara, M. J. Bronikowski, M. Hoenk, B. D. Hunt, and P. H. Siegel, J. Vac. Sci. Technol. B **23**, 157 (2005).

[Mat] Matrox Meteor II, Matrox Electronic Systems GmbH, http://www.matrox.com.

[Mat05] A. Matheisen, "Status of EP at DESY (update),"TTC meeting, Frascati, Italy, (2005).

[Mau06] F. Maurer, A. Dangwal, D. Lysenkov, G. Müller, M. E. Toimil-Molares, C. Trautmann, J. Brötz, and H. Fuess, Nucl. Instr. Meth. Phys. Res. B **245**, 337 (2006).

[Mcc07] D. McClain, J. Wu, N. Tavan, J. Jiao, C. M. McCarter, R. F. Richards, S. Mesarovic, C. D. Richards, and D. F. Bahr, Phys. Chem. C **111**, 7514 (2007).

[Mil04] W. I. Milne, K. B. K. Teo, G. A. J. Amaratunga, P. Legagneux, L. Gangloff, J. P. Schnell, V. Semet, V. Thien Binh, and O. Groening, J. Mater. Chem. **14**, 933 (2004).

[Mic] Micos GmbH, http://www.micos.ws.

[Mil26] R.A. Millikan and C.F. Eyering, Phys. Rev. **27**, 51 (1926).

[Mod84] A. Modinos, "Field, Thermionic, and Secondary Electron Emission Spectroscopy," Plenum Press, New York, ISBN: 0306413213 (1984).

[Mou07] M. S. Mousa, Surf. Interface Anal. **39**, 102 (2007).

[Mül02] T. Müller, K.-H. Heinig, and B. Schmidt, Mater. Sci. Eng. C **19**, 209 (2002).

[Mül98] G. Müller, A. Göhl, T. Habermann, A. Matheisen, D. Nau, D. Proch, and D. Reschke, Proc. of 6th Eur. Part. Acc. Conf. EPAC 98, Stockholm, p. 1876 (1998).

[Mws] MicroWave Studio, CST GmbH.

[Nav08] A. Navitski, G. Müller, T.W. Cornelius, and C. Trautmann, Techn. Digest of the 21st Int. Vacuum Nanoelectronics Conf., Wroclaw, Poland, ISBN: 83-914886-2-4, p. 75-76 and EM5 on CD (2008).

[Nav09a] A. Navitski, G. Müller, V. Sakharuk, T.W. Cornelius, C. Trautmann, and S. Karim, Eur. Phys. J. Appl. Phys. **48**, 30502 (2009).

[Nav09b] A. Navitski, V. Sakharuk, F. Jordan, G. Müller, S. Müller, M. Rauber, M.E. Toimil-Molares, and C. Trautmann, Techn. Digest of the 22nd Int. Vacuum Nanoelectronics Conf., Hamamatsu, IEEE Cat. No. CFP09VAC-PRT, p. 137-138 (2009).

[Nav09c] A. Navitski, G. Müller, K. Flöttmann, and S. Lederer, Proc. of the 14th Int. Conf. on RF Superconductivity, Berlin, TUPPO044, p. 312 (2009).

[Nav09d] A. Navitski, S. Lagotzky, G. Müller, D. Reschke, and X. Singer, Proc. of the 14th Int. Conf. on RF Superconductivity, Berlin, TUPPO045, p. 316 (2009).

[Nav09e] A. Navitski, V. Sakharuk, G. Müller, T.W. Cornelius, C. Trautmann, and S. Karim, GSI Report 2008, p. 356 (2009).

[Nav10a] A. Navitski, G. Müller, V. Sakharuk, A. L. Prudnikava, B. G. Shulitski, and V. A. Labunov, J. Vac. Sci. Technol. B **28**, C2B14 (2010).

[Nav10b] A. Navitski, P. Serbun, B. Bornmann, G. Müller, R. K. Joshi, J. Engstler, and J. J. Schneider, "Observation of hot spots and partial disruptions as field emission limiting factors of pure and TiO_2 coated CNT block arrays," to be published (2010).

[Nav10c] A. Navitski, P. Serbun, G. Müller, J. Engstler, R. K. Joshi, and J. J. Schneider, Proc. of the 23^d Int. Vacuum Nanoelectronics Conf. IVNC2010, Palo Alto, USA, P2-23, p.167 (2010).

[Nav10d] A. Navitski, P. Serbun, G. Muller, I. Alber, M.E. Toimil-Molares, and C. Trautmann, Proc. of the 23^d Int. Vacuum Nanoelectronics Conf. IVNC2010, Palo Alto, USA, P2-21, p. 163 (2010).

[Nav10e] A. Navitski, V. Sakharuk, F. Jordan, G. Müller, I. Alber, M.E. Toimil-Molares, and C. Trautmann, GSI Scientific Report 2009, p. 451 (2010).

[Nav10f] A. Navitski, S. Lagotzky, G. Müller, D. Reschke, X. Singer, Proc. of DPG meeting, Bonn, T.78.7 (2010).

[Nie86] P. Niedermann, Ph. D. Thesis, University Genf, No 2197 (1986).

[Nil00] L. Nilsson, O. Groening, C. Emmenegger, O. Kuettel, E. Schaller, L. Schlapbach, H. Kind, J. M. Bonard, and K. Kern, Appl. Phys. Lett. 76, 2071 (2000).

[Orv89] W. J. Orvis, C. F. McConaghy, D. R. Ciarlo, J. H. Yee, and E.W. Hee, IEEE Trans. on Electron Devices **36**, 2615 (1989).

[Pad09] H. Padamsee, "RF superconductivity," Wiley, ISBN 978-3-527-40572-5.A (2009).

[Pho] Information on the photocathodes is available at http://wwwlasa.mi.infn.it/ttfcathodes/

[Pie28] R. J. Piersol, Phys. Rev. **31**, 441 (1928).

[Pon86] L. Ponto, M. Hein, Elektropolitur von Niob, Technical Report WUP 86-17, Bergische Universität Wuppertal (1986).

[Pro01] D. Proch, D. Reschke, B. Guenther, G. Müller, and D. Werner, Proc. of 10th Workshop on RF Superconductivity, Tsukuba, p. 463 (2001).

[Pru09a] A.L. Prunikava, B.G. Shulitski, V.A. Labunov, A. Navitski, V. Sakharuk, and G. Müller, Techn. Digest of the 22nd Int. Vacuum Nanoelectronics Conf., Hamamatsu, IEEE Cat. No. CFP09VAC-PRTP2-B08, p. 257 (2009).

[Pru09b] A. Prudnikava, V. Labunov, B. Shulitski, V. Sakharuk, A. Navitski, G. Müller, Proc. of the 29th Int. Display Research Conference Eurodisplay 2009, Rome, Italy, p. 319 (2009).

[Pup96] N. Pupeter, Ph. D. Thesis WUB-DIS 96-16, University of Wuppertal (1996).

[Res04] D. Reschke, A. Brinkmann, D. Werner, and G. Müller, Proc. of the LINAC 2004, Lübeck, Germany, THP71, p. 776 (2004).

[Res05a] D. Reschke, 12th International Workshop on RF Superconductivity, Cornell, USA, SuP03 (2005).

[Res05b] D. Reschke, Proceedings of Workshop on Pushing the Limits of RF Superconductivity, Argonne National Laboratory, Report ANL-05/10, Argonne (2005).

[Res07a] D. Reschke, A. Brinkmann, K. Floettmann, D. Klinke, J. Ziegler, D. Werner, R. Grimme, and C. Zorn, Proc. of the 12th Workshop on Rf superconductivity, Beijing, China, TUP48 (2007).

[Res07b] D. Reschke, Proc. of the 2007 Asian Part. Accel. Conf., Indore, India, p. 26 (2007).

[Rin95] G. Rinzler, J. H. Hafner, P. Nikolaev, L. Lou, S. G. Kim, D. Tomanek, P. Nordlander, D. T. Colbert, and R. E. Smalley, Science **269**, 1550 (1995).

[Rot26] F. Rother, Ann. Der Physik **81**, 316 (1926).

[Röt93] R. Röth, Ph.D. Thesis, University of Wuppertal, WUP-DIS 92-12 (1993).

[Sai03] K. Saito, Proc. of the 2003 Part. Accel. Conf. Portland, USA, p. 462 (2003).

[Sai89] K. Saito, Proc. of the 4th Workshop on RF Superconductivity, Tsukuba, p. 63 (1989).

[Sch05] F. Schölz, A. Farr, E.Wappler, M. Mück, A. Brinkmann, and W. Singer, Proc. of 12th Workshop on RF Superconductivity SRF, Ithaca, USA (2005).

[Sch08] S. Schreiber, S. Lederer, P. Michelato, L. Monaco, D. Sertore, and J. H. Han, Proceedings of the 30th FEL08 conf., Gyeongju, Korea, p. 552 (2008).

[Sch10a] R. Schreiner, F. Dams, P. Serbun, A. Navitski, G. Müller, Proc. of the 2nd Int. Workshop on Novel Developments and Applications of Sensor and Actuator Technology, Coburg (2010).

[Sch10b] P. R. Schwoebel, C. E. Holland, and C. A. Spindt, Proc. of the 23^{d} Int. Vacuum Nanoelectronics Conf. IVNC2010, Palo Alto, USA, 3.3, p.14 (2010).

[Sch23] W. Schottky, Z. Physik. **14**, 63 (1923).

[Sch81] K. Schulze, O. Bach, D. Lupton and F. Schreiber, Proceedings of the International Symposium, San Francisco, USA, p.163 (1981).

[Sem02] V. Semet, Vu T. Binh, P. Vincent, D. Guillot, K. B. K. Teo, M. Chhowalla, G. A. J. Amaratunga, W. I. Milne, P. Legagneux, and D. Pribat, Appl. Phys. Lett. **81**, 343 (2002).

[Ser00] D. Sertore, S. Schreiber, K. Flöttmenn, F. Stephan, K. Zapfe and P. Michelato, Nucl. Instr. and Meth. A **445, 422** (2000).

[She90] R. Sherman and W. Whitlock, J. Vac. Sci. Technol. B **8**, 563 (1990).

[Shc62] G. P. Shcherbakov and I. L. Sokol'skaya, "Experimental investigation of energy distribution of field-emitted electrons from CdS single crystals," Sov. Phys. Solid State **4**, 3526 (1962).

[Sin03] W. Singer, A. Brinkmann, D. Proch and X. Singer, Physica C: Superconductivity **386**, 379 (2003).

[Sim67] J. G. Simmons and R. R. Verderber, Appl. Phys. Lett. **10**, 197 (1967).

[Sno09] D. W. Snoke, "Solid State Physics: Essential concepts," ISBN: 0-8053-8664-5, San Francisco (2009).

[Sol] Dassault Systèmes SolidWorks Corp., http://www.solidworks.com.

[Sol08a] D. Solovei, V. Sakharuk, A. Mozalev, A. Navitski, A. Prudnikava, G. Gorokh, G. Müller, Techn. Digest of the 21st Int. Vacuum Nanoelectronics Conf., Wroclaw, Poland, ISBN: 83-914886-2-4, f-ch19, p. 118 (2008).

[Sol08b] D. Solovei, G. Gorokh, A. Mozalev, V. Sakharuk, J. Shaman, A. Navitski, G. Mueller, Proc. of the 1st Int. Sci. Conf. Nano 2008, Minsk, Belarus p. 408 (2008).

[Sol09] D. V. Solovei, V. N. Sakharuk, A. M. Navitski, G. G. Gorokh, G. Müller, Proc. of the 19th Int. Crimean Conf. "Microwave & Telecommunication Technology", Sevastopol, IEEE Cat. No. CFP09788, p. 601 (2009).

[Som33] A. Sommerfeld and H. Bethe, "Handbuch der Physik," **24**, 441 (1933).

[Spi68] C. A. Spindt, J. Appl. Phys. **39**, 3504 (1968).

[Spi76] C.A. Spindt, I. Brodie, L. Humphrey, and E. R. Westerberg, J. Appl. Phys. **47**, 5248 (1976).

[Suh02] J. S. Suh, K. S. Jeong, J. S. Lee, and I. Han, Appl. Phys. Lett., **80**, 2392 (2002).

[Swa67] L.W. Swanson and L.C. Crouser, Phys. Rev. **163**, 622 (1967).

[Teo01] K. B. K. Teo, M. Chhowalla, G. A. J. Amaratunga, W. I. Milne, D. G. Hasko, G. Pirio, P. Legagneux, F. Wyczisk, and D. Pribat, Appl. Phys. Lett. **79**, 1534 (2001).

[Toi01] M. E. Toimil-Molares, J. Brötz, V. Buschmann, D. Dobrev, R. Neumann, R. Scholz, I. U. Schuchert, C. Trautmann, and J. Vetter, Nucl. Instr. Meth. Phys. Res. B **185**, 192 (2001).

[Toi04] M. E. Toimil-Molares, A. G. Balogh, T. W. Cornelius, R. Neumann, C. Trautmann, Appl. Phys. Lett. **85**, 5337 (2004).

[Ttc08] TTC-Report 2008-05 at DESY.

[Tud] Technical university of Darmstadt, TUD, http://www.tu-darmstadt.de.

[Uts91] T. Utsumi, IEEE Transactions on Electronical Devices **38**, 2276 (1991).

[Vil04] L. Vila, P. Vincent, L. Dauginet-De Pra, G. Pirio, E. Minoux, L. Gangloff, S. Demoustier-Champagne, N. Sarazin, E. Ferain, R. Legras, L. Piraux, and P. Legagneux, Nano Lett. **4**, 521 (2004).

[Wan97] J.W. Wang, and G.A. Loew, "Field emission and rf breakdown in high-gradient room-temperature Linac structures," SLAC-PUB-7684 (1997).

[Wen00] H. M.Wen, W. Singer, D. Proch, L. Z. Lin, L.Y. Xiao, Proc. of the 18th International Cryogenic Engineering conference (ICEC18), Mumbai, India (2000).

[Woo97] R.W.Wood Phys. Rev. **5** (1897).

[Xfe] X-ray free electron XFEL, http://www.xfel.eu/de.

[Xih09] C. Xihong, C. Yongqin, W. Wang and Y. Dapeng, Solid State Communications **149**, 523 (2009).

[Xu05] N. S. Xu and S. E. Huq, Material Science Engineering **48**, 47 (2005).

[Xu86] N. S. Xu and R. V. Latham, J. Phys. D: Appl. Phys. **19**, 477 (1986).

[Xu94] N. S. Xu, Y. Tzeng, and R.V. Latham, J. Phys. D: Appl. Phys. **27**, 1988 (1994).

[Xu95] N. S. Xu, "High Voltage Vacuum Insulation," edited by R. V. Latham, Academic Press, London (1995).

[Yeo04] K. S. Yeong, J. T. L. Thong, Applied Surf. Sci. **233**, 20 (2004).

[Yi02] W. K. Yi, T. W. Jeong, S. G. Yu, J. N. Heo, C. S. Lee, J. H. Lee, W. S. Kim, J. B. Yoo, and J. M. Kim, Adv. Mater. **14**, 1464 (2002).

[Yos01] T. Yoshimoto, T. Iwata, S. Kikuchi and N. Yokogawa, Jpn. J. Appl. Phys. **40**, 4197 (2001).

[Yos10] T. Yoshimoto, K. Sato, and T. Iwata, Jpn. J. Appl. Phys. **49**, 070212 (2010).

[Zei88] A. Zeitoun-Fakiris and B. Jütntner, J. Phys. D: Appl. Phys. **21**, 960 (1988).

[Zei91] A. Zeitoun-Fakiris and B. Jütntner, J. Phys. D: Appl. Phys. **24**, 750 (1991).

[Zho02] D. Y. Zhong, G. Y. Zhang, S. Liu, T. Sakurai, and E. G. Wang, Appl. Phys. Lett. **80**, 506 (2002).

[Zhu09] H. Zhua, C. Masarapub, J. Weia, K. Wanga, D. Wua and B. Weib, Physica E: Low-dimensional Systems and Nanostructures, **41**, 1277 (2009).

Appendix A

The user interface of the programs, developed for control of the SAFEM and to perform regulated voltage (field) scans, FE investigations of the localized emitters, and analysis of the data are given in figures A1, A2, and A3.

The program for the regulated voltage (field) scans allows the following:
- selection of the scanning resolution by choosing an appropriate anode and a scanning step (by default, the scanning step corresponds to 1/2 of the anode size);
- selection of the electrode gap;
- setting of the field correction factor α of the anode;
- selection of the scanning range by setting a number of the lines and the points-per-line to scan with the above mentioned step size;
- selection between four scanning modes: line-by-line or meander-type in x or y direction for both. Meander-type mode is faster due to avoiding an idle back move, but it is recommended only for low-resolution scans due to a possible backlash of the sliding tables;
- correction of the tilt between the anode moving plane and the cathode surface;
- selection of an interesting area to scan;
- calibration of the sliding stages;
- setting of the required velocity, acceleration, etc. of the sliding stages;
- monitoring of the actual voltage and current during the scans. Regulation of the voltage, by the way, is done via the analog feedback system with PID-regulation, which couples the ammeter and the power supply;
- imaging of field and voltage maps;
- automatic storage of the data;
- creation of a report containing all important initial parameters and resulted maps in html format.

The program for the FE investigations of the localized emitters allows the following:
- selection of emitters on the basis of the previous scans;
- determination of the local distance d (requires the program shown in Fig. A3);
- I-V measurements by manual voltage rise/fall and automatic acquisition of the current from the picoammeter. It is planned to add also an automatic voltage control as in the FESM;

- imaging of I-V, I-E, I-t, U-t, and FN plots;
- determination of the onset field and calculation of the β and S factors;
- automatic storage of the data;
- creation of a report containing all important initial parameters, measured data and plots in html format.

For late analysis of the stored raw data of the scans and I-V measurements, two additional programs shown in Figs. A4, A5 are foreseen. Few examples of additional programs required for the SAFEM operation are shown in figures A6 and A7.

Appendix

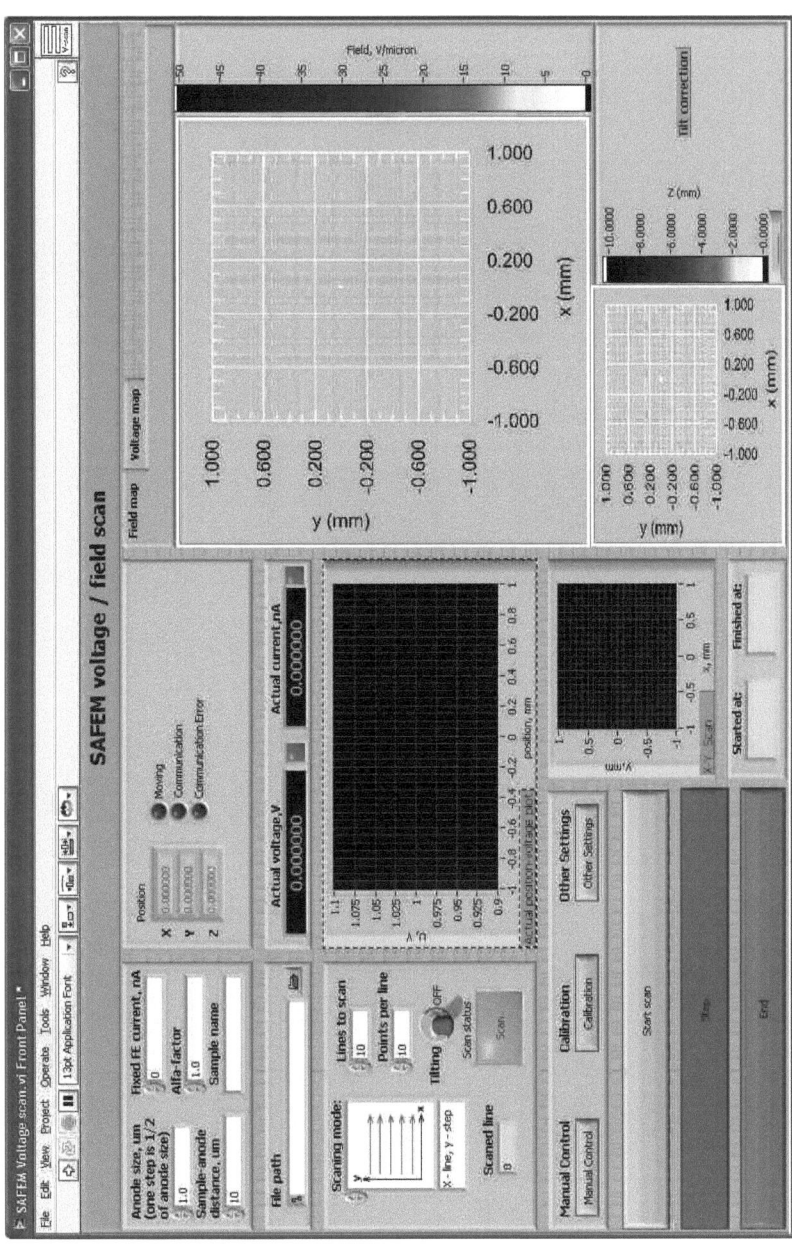

Fig. A1: User interface of the program for regulated voltage (field) scans.

Appendix

Fig. A2: *User interface of the program for the FE investigations of individual emitters.*

Appendix

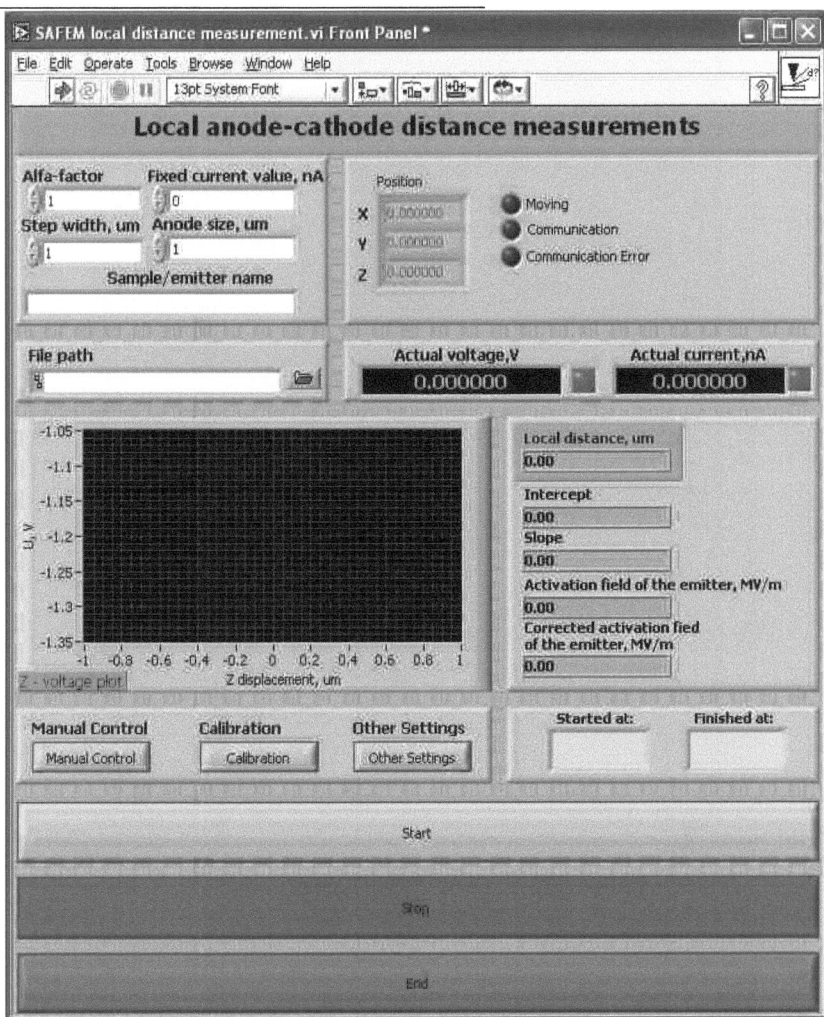

Fig. A3: User interface of the program for the local distance measurements.

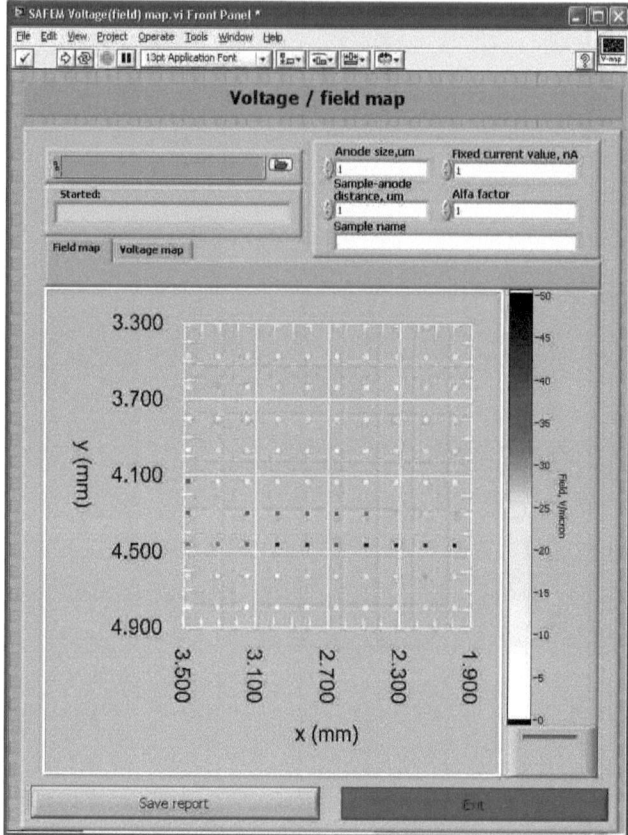

Fig. A4: User interface of the program for analysis of the voltage (field) maps.

Appendix

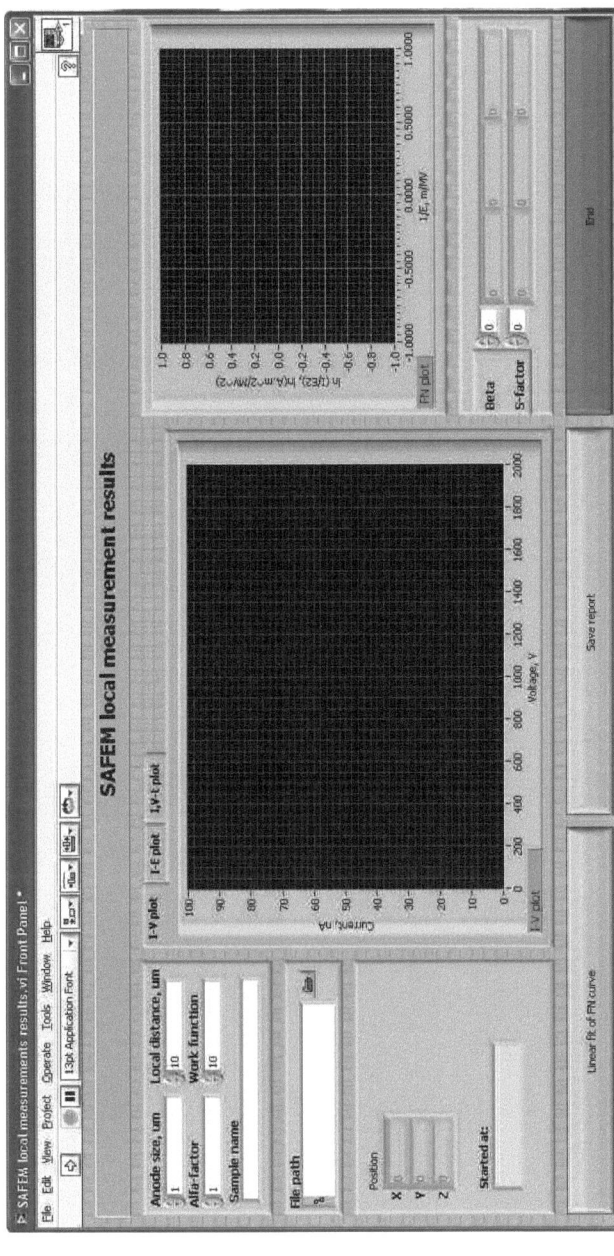

Fig. A5: User interface of the program for analysis of the I-V data.

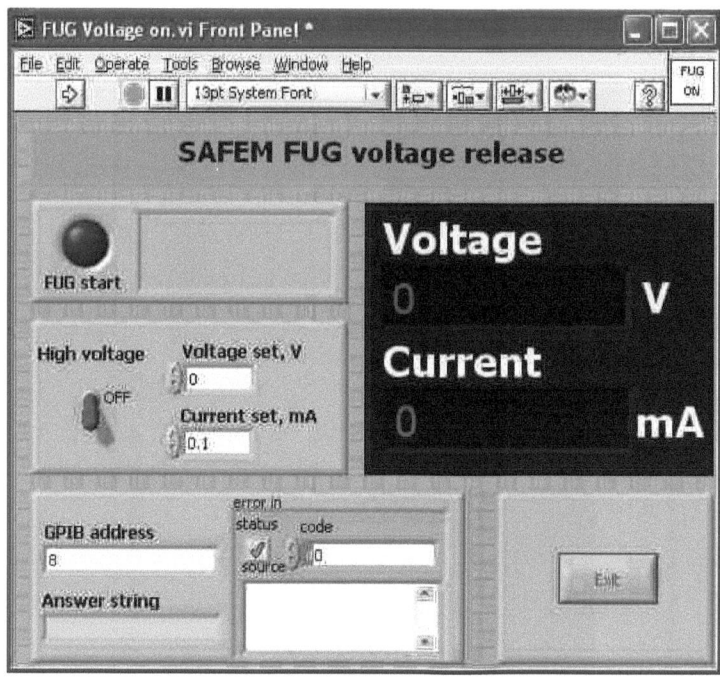

Fig. A6: User interface of the program for control of the FUG power supply.

Appendix 125

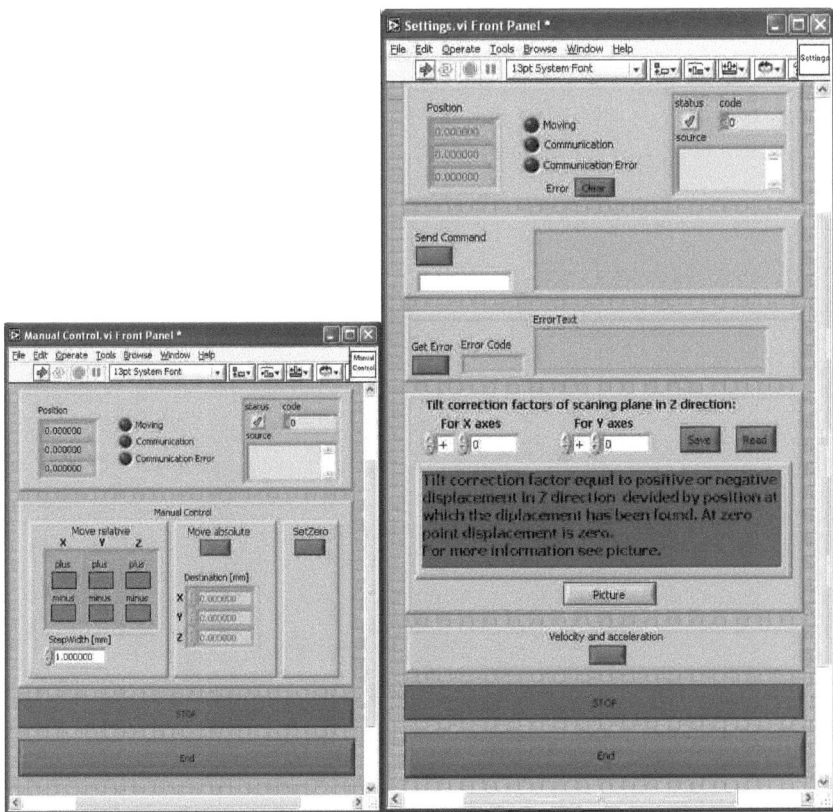

Fig. A7: User interface of the programs for the manual control of the sliding stages and the tilt correction.

I want morebooks!

Buy your books fast and straightforward online - at one of world's fastest growing online book stores! Environmentally sound due to Print-on-Demand technologies.

Buy your books online at
www.morebooks.shop

Kaufen Sie Ihre Bücher schnell und unkompliziert online – auf einer der am schnellsten wachsenden Buchhandelsplattformen weltweit! Dank Print-On-Demand umwelt- und ressourcenschonend produziert.

Bücher schneller online kaufen
www.morebooks.shop

KS OmniScriptum Publishing
Brivibas gatve 197
LV-1039 Riga, Latvia
Telefax: +371 686 204 55

info@omniscriptum.com
www.omniscriptum.com

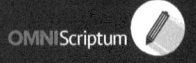

Printed by Books on Demand GmbH, Norderstedt / Germany